Developments in Materials Characterization Technologies

Symposium held 23 and 24 July 1995, during the 28th Annual Technical Meeting of the International Metallographic Society, Albuquerque, New Mexico, USA

Edited by
George F. Vander Voort, *Carpenter Technology Corporation*
John J. Friel, *Princeton Gamma-Tech, Inc.*

Veronica Flint, *Manager Book Acquisitions*
Grace M. Davidson, *Manager Book Production*
Cheryl L. Powers, *Production Project Coordinator*

International
Metallographic
Society

The International Metallographic Society
Columbus, Ohio, USA

The Materials
Information Society

ASM International
Materials Park, Ohio, USA

Library of Congress Catalog Card Number: 96-84817
ISBN: 0-87170-580-X
SAN: 204-7586

ASM International®
Materials Park, OH 44073-0002
Printed in the United States of America

CONTRIBUTORS

Mel D. Ball
Alcan International Ltd.
Kingston Research & Development Centre
Box 8400
Kingston, Ontario, Canada K7L 5L9
613-541-2038
FAX: 613-541-2134

John A. Brooks
Sandia National Laboratory
P.O. Box 969, MS 9403
Livermore, CA 94550
510-294-2051
FAX: 510-294-3410

Mary Grace Burke (ZAP 13A)
Westinghouse Electric Corp.
Bettis Atomic Power Lab
P.O. Box 79
West Mifflin, PA 15122
412-476-5883
FAX: 412-476-5151

Raymond W. Carpenter
Arizona State University
Center for Solid-State Science
Tempe, AZ 85287-1704
602-965-4549
FAX: 602-965-9004

John J. Friel
Princeton Gamma-Tech, Inc.
1200 State Road
Princeton, NJ 08540
609-924-7310
FAX: 609-924-1729

Arun M. Gokhale
Georgia Institute of Technology
School of Materials Science
Atlanta, GA 30332-0245
404-894-2887
FAX: 404-853-9140

Joseph R. Michaels
Sandia National Labs
Organization 1822
P.O. Box 5800
Albuquerque, NM 87185-0342
505-844-9115
FAX: 505-846-4168

Lawrence E. Murr
The University of Texas at El Paso
Dept. of Metallurgy and Materials Science
El Paso, TX 79968-0520
915-747-6929
FAX: 915-747-5616

Rod H. Packwood
CANMET
Metals Technology Laboratories
568 Booth Street
Ottawa, Ontario, Canada K1A 0G1
613-992-2288
FAX: 613-992-8735

Paul S. Prevey
Lambda Research
5521 Fair Lane
Cincinnati, OH 45227
513-561-0883
FAX: 513-561-0886

James H. Richardson
The Aerospace Corp.
P.O. Box 92957
MS M4/994
Los Angeles, CA 90009
310-336-5439
FAX: 310-336-6914

Brian R. Strohmeier
Alcoa Technical Center
Surface Technology Division
100 Technical Drive
Alcoa Center, PA 15069-0001
412-337-2533
FAX: 412-337-2809

George F. Vander Voort
Carpenter technology Corp.
R & D Center
P.O. Box 14662
Reading, PA 19612-4662
610-208-2766
FAX: 610-208-3256

PREFACE

One of the featured events of the 28th Annual International Metallographic Society Convention, held in Albuquerque, New Mexico, on 23-26 July 1995, was the symposium *Developments in Materials Characterization Technologies*. The convention, chaired by E. Daniel Albrecht, was held at the Hyatt Regency Albuquerque. This volume is a collection of thirteen of the presentations made at the Symposium.

Lawrence Murr, of the University of Texas at the El Paso and a regular contributor at IMS meetings, kicked off the symposium with a review of several materials characterization case histories requiring the combined application of a number of techniques - LOM, SEM, TEM, Auger, and so forth - to provide the necessary information. Application of these complementary techniques was referred to as "multi-dimensional" analysis.

Be it ever so humble, the light optical microscope is still the cornerstone of any materials characterization effort. Although the light microscope may be considered as a mature instrument, Jim Richardson of the Aerospace Corporation and a noted author on the subject, reviewed many of the important refinements that have occurred over the past two decades, refinements that have enhanced optical images, operator convenience and versatility. The large console metallograph may be a dinosaur but the new line of less expensive upright and inverted compact metallographs offer improved performance at less cost. Jim also reviewed the newer areas of electronic image enhancement and documentation - new tricks for an old technology.

The incredible growth of personal computers and digital imaging technology has stimulated development of powerful image analyzers. George Vander Voort of Carpenter Technology Corporation reviewed three ASTM image analysis-based standard test methods for characterizing materials. Weaknesses associated with E-45 chart-based inclusion ratings can be overcome by using the E-1122 image-analysis method. E-1245, a stereo logical image analysis method for characterizing second-phase constituents is suitable for computer databases. E-1382 covers grain size measurements performed by image analyzers.

Arun Gokhale of the Georgia Institute of Technology presented some new ideas for characterizing the spatial arrangement of fibers in unidirectional fiber composites with the aid of image analysis. The microstructure was characterized by fiber-rich and fiber-poor regions. The radial distribution function was employed to describe clustering in the fiber-rich areas and the scarcity of fibers in the fiber-poor regions.

Since its commercial introduction just over thirty years ago, the scanning electron microscope (SEM) has become one of the most powerful, and desirable, tools in the metallography laboratory. Mel Ball of Alcan International Limited presented an historical review of its development and an overview of the many sources of structural, compositional and crystallographic information obtainable.

Energy-dispersive spectroscopy (EDS) developed along with the SEM, making it a much more valuable instrument. John Friel, Princeton Gamma-Tech, Inc., reviewed the history of EDS development illustrating peak resolution of the same compound at different time periods.

Back scattered electron Kikuchi patterns generated with the SEM provide a means for phase identification and crystal orientation. Joe Michael of Sandia National Laboratories (Albuquerque, NM) presented a review of this technique, along with his recent improvements, and practical examples illustrating its application.

The electron microprobe analyzer (EMPA) has been a very powerful research tool for quantifying chemistry in very small regions. Wavelength-dispersive spectroscopy offers greater sensitivity and precision than energy-dispersive spectroscopy. Rod Packwood described the analytical approaches used to convert x-ray counts to chemical analysis results and presented several classical analytical problems and their EMPA solutions.

With the development of the SEM and EDS, it was a natural extension to apply this technology to the transmission electron microscope, considered to be a "mature" instrument twenty-five years ago. Further, the beam size was reduced converting electron diffraction to the more powerful convergent beam electron diffraction procedure for compound identification and crystal orientation determination. EDS and electron energy loss spectroscopy were added to provide chemistry information. The development and transformation of the TEM to the AEM (analytical electron microscope) was reviewed by Ray Carpenter from Arizona State University.

Analytical electron microscopy has become a very powerful tool. When coupled with atom probe field ion microscopy. This capability is further enhanced, as described by Mary Grace Burke from the Bettis Atomic Power Laboratory of the Westinghouse Electric Corporation. This combination extends analysis down to regions as small as 1 nm.

Characterization of residual stresses can be extremely important in failure analysis studies and in efforts to improve product performance. While there are a number of techniques for measuring residual stresses, the x-ray diffraction method is one of the most powerful and most useful. Paul Prevey of Lambda Research described the theory behind XRD residual stress measurements and provided examples of its application.

While metallurgists have devoted much time and effort to quantifying chemistry in ever smaller regions, little has been done to map larger scale chemistry variations. John Krafcik and John Brooks of the Sandia National Laboratories (Livermore, CA) described the use of energy-dispersive spectroscopy to map chemistry over large areas in as-cast vacuum arc remelted superalloys.

Over the past twenty years, Auger electron spectroscopy has become the most popular tool for characterizing the surface composition of materials. Brian Strohmeier, Alcoa Technical Center, described the history of Auger development and the unique aspects of Auger electrons that make them so useful for characterizing surface chemistry. Recent developments that extend the usefulness of AES were also presented.

George F. Vander Voort
Carpenter Technology Corp.
Reading, Pennsylvania

John J. Friel
Princeton Gamma-Tech, Inc.
Princeton, New Jersey

CONTENTS

Multi-Dimensional Microanalysis in Materials Characterization: Some Case Examples

**Lawrence E. Murr, C-S. Niou, E. Ferreyra T., S. Pappu, C. Kennedy, S.A. Quinones,
J. M. Rivas, R. J. Romero and J. G. Maldonado**
The University of Texas at El Paso, El Paso, TX

Abstract

Multi-dimensional analysis involves the necessity to utilize several analytical tools or microanalysis approaches to resolve or identify microstructural and microchemical issues in materials: optical metallography, SEM, TEM, Auger and other spectroscopies and microscopies involving light, acoustic signals, electron and ion beams, and X-rays. This paper illustrates these issues with four case examples: polycrystalline copper; deformation twinning; hypervelocity impact craters; $M_{23}C_6$ precipitation in stainless steel. TEM emerges as possibly the most powerful tool for a broad range of materials characterization.

THE CONCEPT OF MULTI-DIMENSIONAL ANALYSIS or microanalysis is not new. In fact Murr (1,2) has developed the theme of integrated modular micro-analysis and other variances of the concept of materials characterization utilizing a multiplicity of microstructural and micro-chemical approaches to understanding structural phenomena and structure-property relationships. Here we will utilize the concept of multidimensional microanalysis to mean the use of various analytical tools and approaches to understand, primarily, structural and micro-structural issues in materials. For the most part, we will be concerned with illustrating how essential it is not to rely upon a single mode of analysis or observation to draw structural and microstructural conclusions. Often the observations are ambiguous or superficial at best. This is particularly true of metallographic (or light microscope) observations of surface structure. All too often crystallographic or microstructural inferences or conclusions are drawn which are unsupported and in fact incorrect. Often these inferences imply microchemical details which are nonexistent in a metallographic image. While we may not want to believe that a considerable amount of scientific research involving materials does not include systematic materials characterization (which is now well established (1,2)), this is indeed a fact, and all too often unsupported microstructural conclusions are perpetuated as fact for decades in the literature.

In this paper, we will consider "multi-dimensional" to represent the spatial concept of dimensionality and the combination of crystallographic and compositional information in materials characterization. This will often involve examination of a material utilizing a variety of excitation or imaging sources and detectors. We will draw upon some relevant case examples not only to illustrate these concepts, but also to provide some new insight into several practical and contemporary materials issues.

Case Examples

Grain Sizes and Grain Size Effects are often the simplest and most apparent micro-structural issues in dealing with materials, particularly metals and alloys. A great deal of quality control in metal manufacturing arenas involves specification of grain size, and corresponding hardness values, usually macroscopic hardness measurements involving Rockwell or Brinell. However, all too often grain sizes and grain structure are not known.

There is also the popular belief that in general, grain size is related to hardness (H) or yield stress (σ_y) through the so-called Hall-Petch relationship: $\sigma_y = \sigma'_o + KD^{-1/2}$; where σ'_o is the interior grain (or matrix) friction stress, K is a constant, and D is the average grain size. It is often implied that the friction stress, σ'_o, represents the yield stress for a perfect single crystal (where D $\rightarrow \infty$). However, Wang and Murr (3) have discussed the fact that solute atoms, other impurities, or dislocations could significantly modify this term so that in reality, $\sigma'_o, = \sigma_o + f(\varepsilon)$; where $(f(\varepsilon)$ represents these intrinsic microstructural features. In addition, industrial processing is often dependent upon the total stored energy per unit volume, which may be correspondingly expressed by $(\Gamma\gamma_{gb}/D+\alpha Gb^2\rho)$; where it can be seen that these two terms represent the grain boundary contribution and the matrix contribution expressed as $f(\varepsilon)$ above: Γ is a geometrical factor, γ_{gb} is the average grain boundary free energy, α is a constant, G is the shear modulus, b is the dislocation Burgers vector, and ρ is the dislocation density.

Figure 1 provides a simple example of the issues discussed above for polycrystalline, OFHC (99.99%) copper. The light microscope images of grain structure in Fig. 1(a) and (c) are essentially identical although there is a slight increase in the average grain size (D) in Fig. 1(c) from Fig. 1(a) (45 μm from 35 μm). However, the Rockwell B hardness and Vickers microhardness (100 gf load) values for Fig. 1(a) are 90 and 74 respectively, while for Fig. 1(c) they are 39 and 59 respectively. This difference in microhardness is primarily due to the stored energy as dislocation cells in Fig. 1(b) in contrast to only remnants of dislocation cells in Fig. 1(d). Correspondingly, the Rockwell B and Vickers hardness values from the well-annealed (low dislocation density) and large-grained (310 μm) copper sample shown in Fig. 1(e) and (f) are 37 and 60, and not much changed from the smaller grained, annealed material. Figure 1 illustrates that it is imperative to peruse the microstructure in polycrystalline materials in order to know whether dislocation substructures or other microstructure contribute significantly to the stored energy.

Deformation Twinning in both fcc and bcc metals and alloys has been an interesting issue for decades, and provides another case example where not only transmission electron microscopy (TEM) is required to actually see twinning features and geometries, but also corresponding selected-area electron diffraction (SAD) patterns are necessary to uniquely identify both twinning reflections and crystallographic directions. A particular case in point involves deformation twinning in bcc tantalum. Originally, twinning (mechanical) was reported in large-grain (~4mm) tantalum hammered at liquid nitrogen temperature (4) then at room temperature (5), and in shock loading (6). But in each of these examples there were only straight, serrated features observed by light microscopy in polished specimens, and in the shock loading case there was nothing known about the pressure or the shock conditions. More recently, Wittman, et al (7) have discussed deformation twinning in tantalum after plane-wave shock loading at ~24 GPa peak pressure utilizing both LM and TEM. But only tiny, dark bands were observed in a TEM image, and no SAD evidence was presented.

Figure 2 shows a tantalum specimen having equiaxed grains roughly 56 μm in diameter subjected to a plane shock pulse of 45 GPa (in 2 μs). The LM image in Fig. 2 shows two adjoining (100) grains each containing linear bands in 4 different <042> directions corresponding to traces of {112} twinning planes. Because there are 4 different crystallographic directions as illustrated in the corresponding diffraction net in Fig. 2 (lower left), the grains can be uniquely determined to be (100). Images of these deformation twins in a (100) orientation are shown in the TEM bright-field image insert in Fig. 2. The corresponding SAD pattern is shown as an insert in the TEM image in Fig. 2. The SAD pattern insert of Fig. 2 shows no twin reflections because the twins in the TEM image, while numerous, are very thin; and the corresponding twin volume is small.

Figure 3 shows twins in the shock-loaded tantalum in a (311) grain surface orientation along with an SAD pattern showing very intense twin reflections corresponding to <112>/3 reflection layers. Fig. 3(c) illustrates the corresponding dark-field TEM image utilizing three unique twin reflections shown circled in the SAD pattern, and corresponding to the {112} plane traces illustrated in Fig. 3(a). Figure 3 provides unambiguous proof for twinning in shock-loaded tantalum and affirms the inferences described for Fig. 2.

Figure 4(a) shows for comparison shock-induced deformation twins in nickel for the same conditions (~45 GPa in 2μs) producing deformation twins in tantalum in Fig. 2. The grain

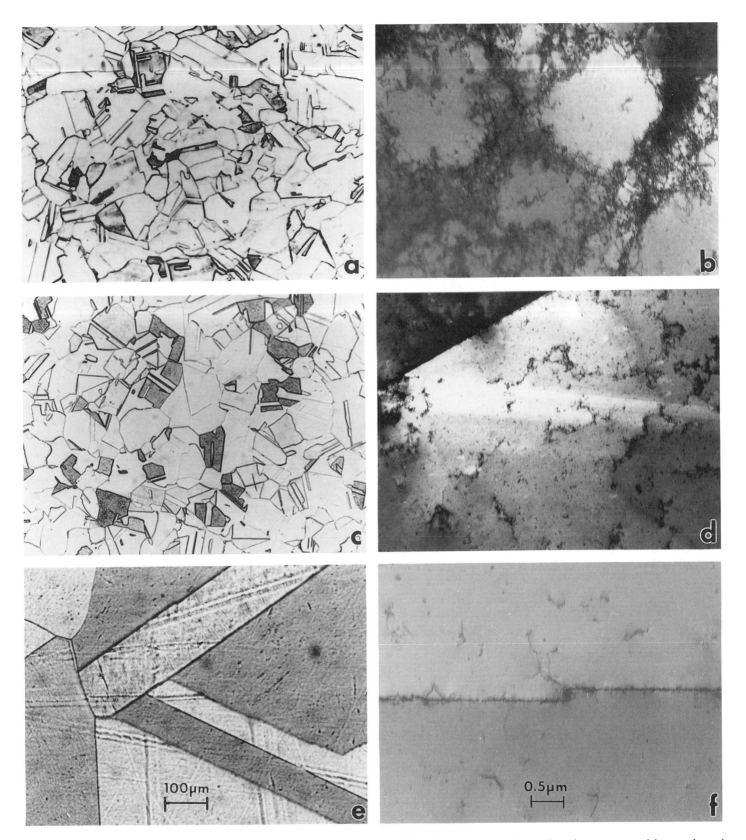

Fig. 1 - Substructures in polycrystalline copper plate. (a) Mill-rolled copper plate showing reasonably equiaxed grain structure by light microscopy. (b) TEM bright-field image showing dislocation cell structure in (a). (c) Sample (a) annealed 2h at 500°C showing slight grain growth. (d) TEM image of (c) showing only remnants of dislocation cell structure. (e) Large grain structure in copper plate after annealing for 10h at 950°C. (f) TEM image showing relatively low dislocation density in (e).

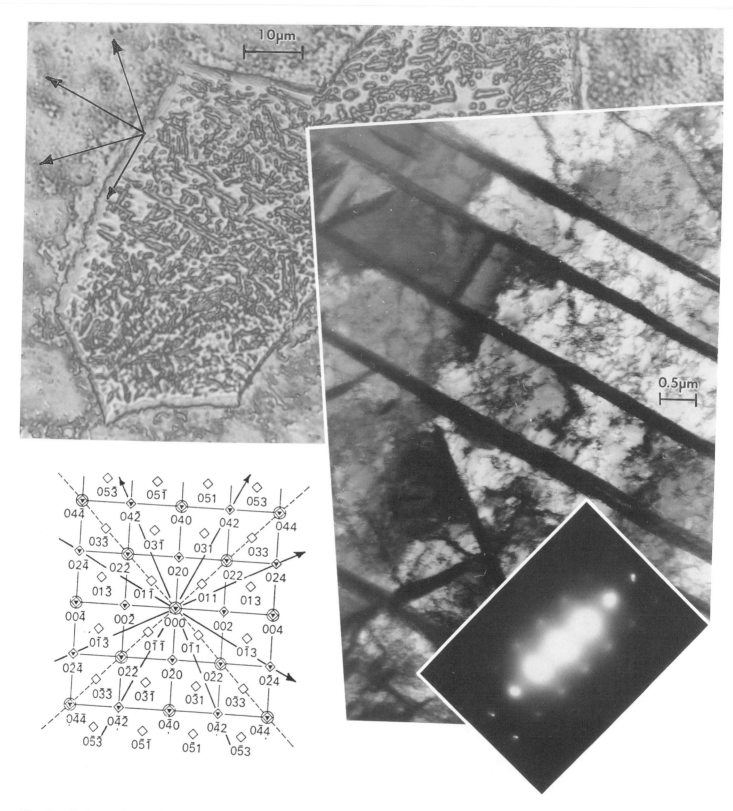

Fig. 2 - Deformation twins in shock loaded tantalum. Upper left shows polished and etched LM surface section with 4 different trace directions. A corresponding TEM (bright-field) image superimposed to the right coincides with the LM image and illustrates the 4 different {112} planes for a (100) grain surface orientation shown by the SAD pattern insert. The (100) diffraction net shown lower left illustrates the corresponding crystallographic indexing. Note this pattern represents bcc (open diamonds), fcc (solid triangles), and diamond cubic (open circles). The 4 trace directions in the (100) bcc tantalum orientation are noted in the diffraction net by arrows. All of the {112} twin planes make an angle of roughly 66° with the (100) grain surface.

4

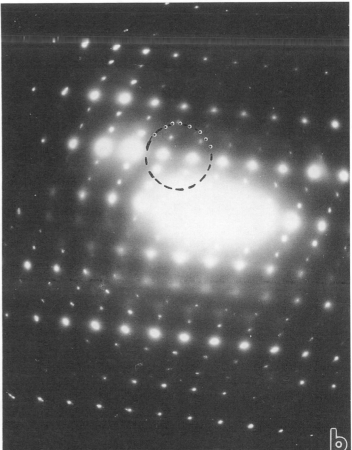

Fig. 3 - Deformation twins in shock loaded tantalum for a (311) grain surface orientation. (a) Bright-field TEM image. (b) SAD pattern coincident with image in (a) showing prominent twin spots in n< $\overline{1}$12 >/3 layers. (c) Dark-field image of (a) using reflections circled in (b), which include all four {112} twin planes. Two twin planes lie in [0$\overline{1}$1] direction and their twin reflections coincide with the corresponding matrix reflections.

surface orientation is (100) and twins lie in {111} planes at ~55° to the grain surface, along <022> directions shown dotted in the indexed diffraction net of Fig. 2. Figure 4(b) shows the corresponding SAD pattern for Fig. 4(a) to exhibit intense <020>/3 twin reflections in layers similar to Fig. 3(b) for tantalum (2). Figure 4 (c) shows deformation twins in molybdenum shock loaded at 26 GPa (~8 μs) in a (100) grain. Unlike the twins in (100) tantalum shown in Fig. 2, only two different {112} planes are twinned. The corresponding SAD pattern in Fig. 4(d) shows the

5

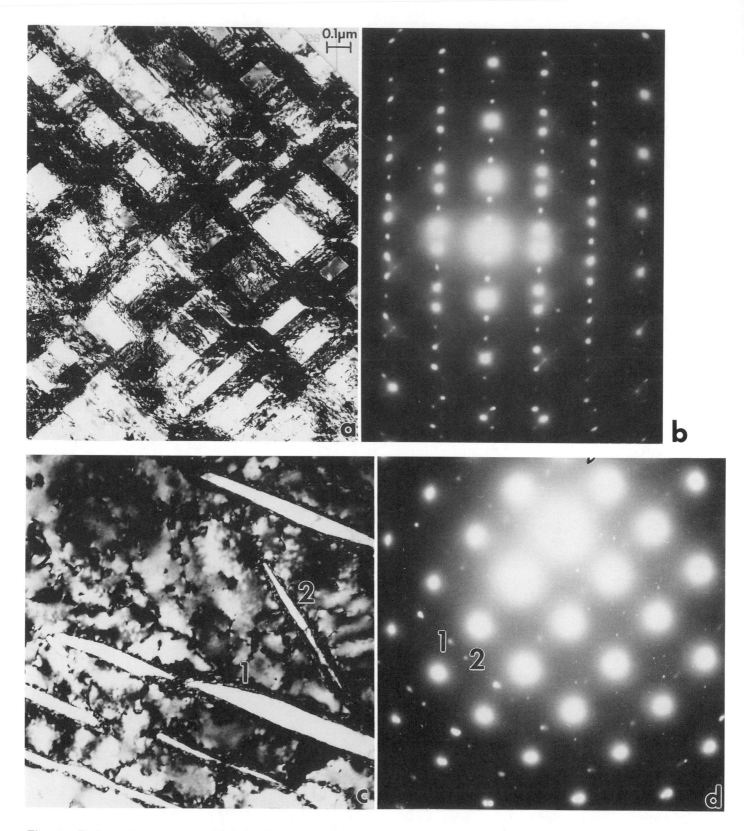

Fig. 4 - Deformation twins in shock loaded, fcc (100) nickel (a) and bcc (100) molybdenum (c). The corresponding fcc SAD pattern in (b) shows prominent twin reflections in n<020>/3 layers. (refer to Fig. 1 diffraction net) These layers are alternately shifted with the matrix reflections. Faint double diffraction spots are also often visible in the inter-layer. (c) Shows two directions of twins corresponding to [0$\bar{2}$4] and [0$\bar{4}$2] in the bcc diffraction net in Fig. 1. (d) Shows the corresponding (100) SAD pattern for (c) showing the two twin reflections corresponding to the two twin trace directions in (c).

6

two twin reflections for these two twinning directions to lie in alternating layers of <011>/3 reflection spots. Deformation twins in molybdenum have been treated more extensively in a paper by Wongwiwat and Murr (8).

Hypervelocity Impact Cratering phenomena has been studied rather intensively for the past 40 years. This has included not only NASA-related impact testing, but also a host of military impact and penetration studies world-wide. Projectile impact velocities have exceeded 10 km/s in laboratory gun experiments. More recently, metal targets exposed in space in low-earth orbit have allowed a wide range of craters formed by very high velocity impacts (>20 km/s) to be examined (9-11)). Figure 5 illustrates some typical features of hypervelocity impact craters observed by a variety of techniques: light microscopy (LM) acoustic microscopy (AM), and scanning electron microscopy (SEM). The limitations of both LM and AM in providing crater details is quite obvious. However, there are also limitations of SEM analysis, and accurate crater geometry measurements really require half-crater section views as shown in Fig. 5(e). Murr and Rivas (12) have in fact explored the prospects of determining projectile impact velocity on the basis of normal (in-plane) crater observations as shown in Fig. 5(c), but such prospects are ambiguous and unreliable. Nonetheless, measurements of crater diameter or crater depth/diameter ratio versus laboratory projectile impact velocity have proven interesting (13). Indeed, dozens of empirical equations have been developed to calculate crater depth and depth/diameter ratios based on assumptions about projectile velocity, and efforts have been made to estimate projectile velocities. However, each must be fitted to specific circumstances. In addition, efforts to relate the target properties to cratering have included macro-hardness parameters and target strength parameters, but there have been no attempts to relate target microstructure, including grain structure, to cratering phenomena. In this regard, there have been few if any attempts to characterize target microstructures, including grain size, and in the recent LDEF experiment (9), not a single target was characterized prior to deployment on orbit; consequently there were no real materials controls.

While Fig. 5 illustrates a multiplicity of crater views and associated observational techniques, there is little information about how the target material responds to the high velocity penetrating projectile. Popular views of crater formation discuss melt interactions and vaporization of the projectile at hypervelocity, and liquid jetting to form the crater rims during crater formation. However little of this phenomena is contained within computer codes which often present reasonable crater formation sequences.

Rivas, et al (14) demonstrated from TEM thin section observations that for a small (~ 1mm) diameter hypervelocity crater in a stainless steel bolt on LDEF, deformation twins occurred at 1 crater diameter from the crater wall (bottom), suggesting that shock wave effects may be important in altering the microstructures associated with hypervelocity impact craters. More recently, it has been shown by TEM analysis of `crater cross-sections that previously unpredicted microstructures occur below hypervelocity impact craters (11).

The copper crater in Fig. 6 exhibits a zone of dynamically recrystallized and recovered material which extends into a wide zone of microbands (Fig. 6(c) which are believed to be related to the shock wave. Indeed, we would have expected this zone to consist of deformation twins similar to those shown for shock-loaded nickel in Fig. 4(a) since the critical twinning pressure (15) for copper is around 20 GPa, and the calculated Bernoulli pressure for the copper crater in Fig. 6(a) is about 25 GPa. In fact, it is interesting to note that one might readily conclude from the LM image in Fig. 6(c) that these linear features were deformation twins. Here again, TEM observations were critical in correctly and unambiguously identifying the microstructural issues. Figure 6(e) illustrates the general features to be expected in connection with shock-induced microstructural changes in a zone below the crater wall. This zone should propagate outward with impact velocity (and pressure).

Figure 7(b) shows the axial microhardness for the copper crater in Fig. 6 in contrast to the computer-simulated strain from Fig. 7(a) (which can be related to residual microhardness through Hooke's law: $\varepsilon = \sigma/E$). Figure 7(c) also alludes to the fact that there remains a great deal of fundamental uncertainty with regard to the pressure calculation and pressure effects associated with hypervelocity projectile impact. For example, the Hugoniot or shock pressure (P_s) calculated at the point of impact (16) is many

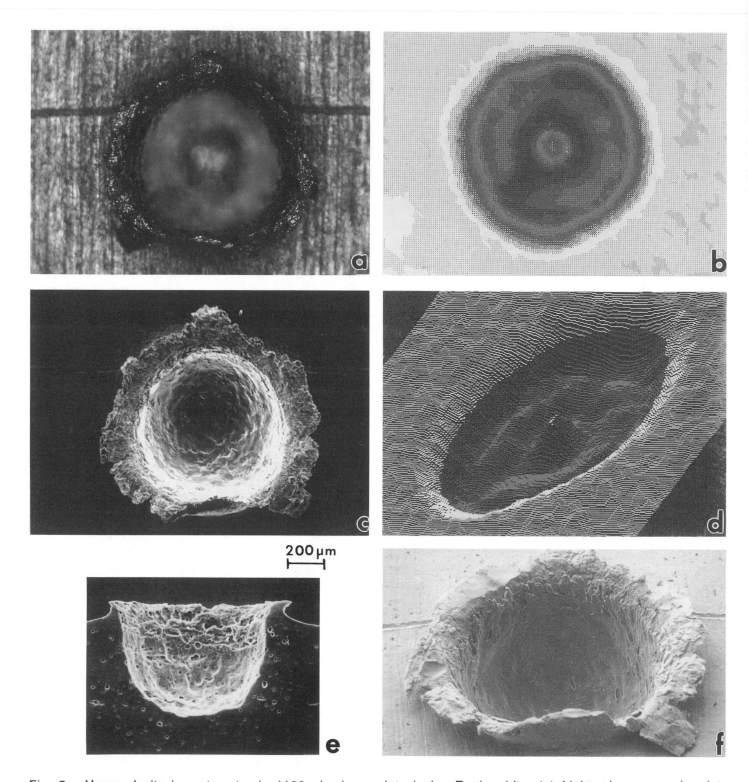

Fig. 5 - Hypervelocity impact crater in 1100 aluminum plate in low-Earth orbit. (a) Light microscope view into crater. (b) Scanning acoustic microscope view into crater. (c) SEM view of crater. (d) Reconstruction of crater geometry from acoustic data in (b). (e) SEM section view of crater. (f) Tilted SEM view of crater in (c).

times larger than the Bernoulli pressure (P_B) which is strictly a fluid dynamic assumption generally associated with the forming craters. It is also very apparent from Fig 7(a) that computer simulations provide no relevant information about the residual crater-related microstructure. For example, compare Fig. 7(a) and (b) with Fig. 6. It may be possible to resolve these issues by very detailed and systematic TEM microstructure mappings around and beyond hypervelocity impact

8

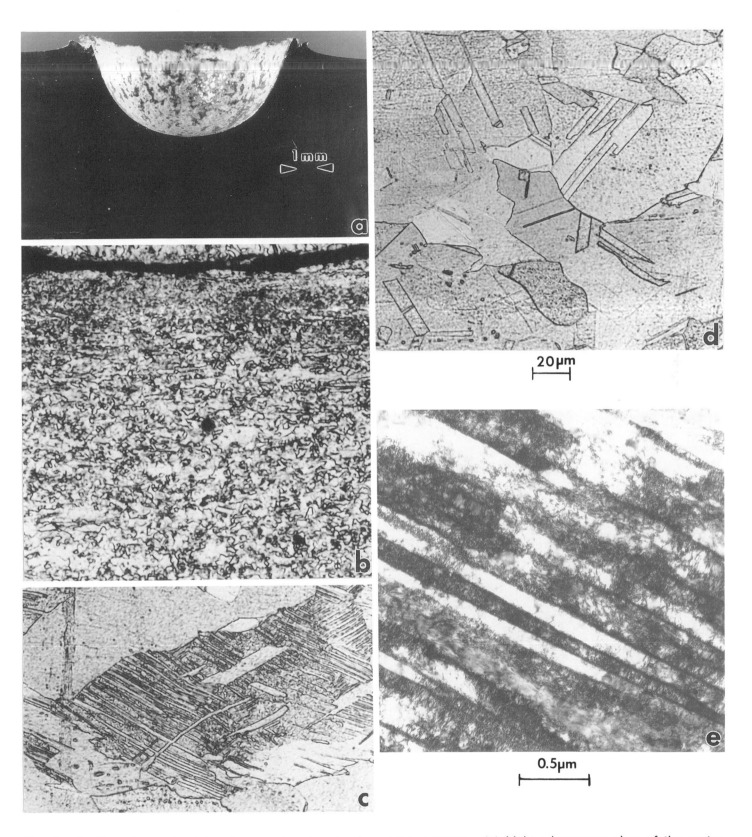

Fig. 6 - Section views of a laboratory-produced crater in a copper target. (a) Light microscope view of the crater produced by 3.2 mm diameter aluminum projectile at 6.7 km/s. (b) LM image of microstructure just below the crater bottom along the impact axis. (c) Typical microstructure roughly 2 mm below the crater bottom showing linear bands. (d) Typical LM view of the original copper target microstructure. (e) TEM image showing the linear bands in (c) to be microbands. The grain surface orientation is (112) and the microbands lie along the [1$\bar{1}$0] direction, coincident with the trace of {111} planes.

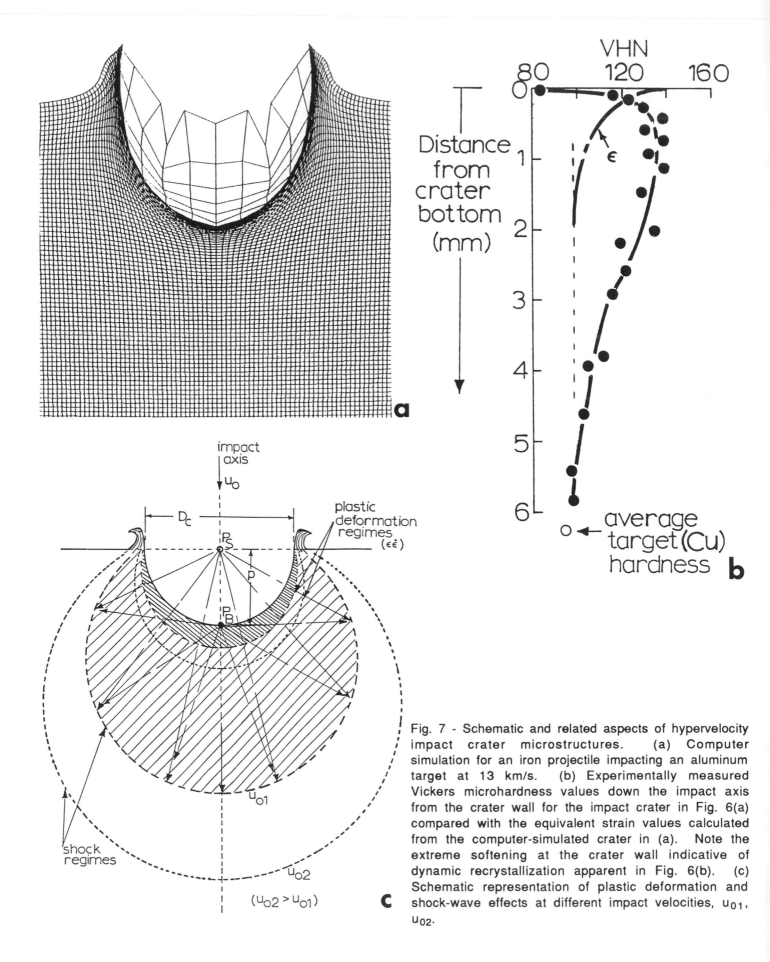

Fig. 7 - Schematic and related aspects of hypervelocity impact crater microstructures. (a) Computer simulation for an iron projectile impacting an aluminum target at 13 km/s. (b) Experimentally measured Vickers microhardness values down the impact axis from the crater wall for the impact crater in Fig. 6(a) compared with the equivalent strain values calculated from the computer-simulated crater in (a). Note the extreme softening at the crater wall indicative of dynamic recrystallization apparent in Fig. 6(b). (c) Schematic representation of plastic deformation and shock-wave effects at different impact velocities, u_{01}, u_{02}.

craters. Microhardness mappings may also be helpful in confirming microstructural issues such as recrystallization where significant softening would be expected as illustrated on comparing Figs. 6(b) and 7(b). Somehow the microstructural issues need to be encoded in the computer simulations as well.

M$_{23}$C$_6$ Precipitates in Stainless Steel will serve as the final case example. Although these precipitates have been observed for decades, there are issues which remain unresolved. Romero and Murr [17] have recently demonstrated, for example, that lamellar M$_{23}$C$_6$ precipitates grow in preferred and consistent crystallographic directions which are not only coincident with {111} planes, but also consistent with a configurational torque theory developed two decades ago [18]. Figure 8 illustrates this specific lamellar growth which requires TEM and SAD to unambiguously demonstrate these features in 304 stainless steel. The intriguing feature of examples shown typically in Fig. 8 is the fundamental role that interfacial or configurational torques may play in the growth of M$_{23}$C$_6$. This in itself is a very interesting scientific issue because, while the torque concept has been discussed a few times for polycrystalline equilibrium phenomena, there have been no known examples of torques influencing specific microstructural phenomena or structure-property relationships.

In some related studies of the site-specific nature of M$_{23}$C$_6$ nucleation and growth, we have also recently examined precipitation on coherent twin boundaries in relation to other interfaces [19]. These issues are related to sensitization phenomena, which for 304 stainless steel are also related to deformation-induced martensite (α'), and some interesting recrystallization phenomena when heavily deformed 304 stainless steel is aged, even for short times. In these cases, the ordinary sensitization features observed in LM (such as precipitation on boundaries) become completely obscure, as illustrated in Fig. 9(a) and (b). However, TEM shows the obscure etching features to result by recrystallization producing an extremely fine, 2-phase microstructure of fcc (γ) and bcc (α') shown in Fig. 9(c). These fine grain structures provide rapid sensitization kinetics and M$_{23}$C$_6$ carbides rapidly nucleate and grow on these interfaces which have interfacial free energies nearly the same as the fcc $\gamma\gamma$ (grain boundary) interface [20]. These features are illustrated by microbeam (or convergent beam) electron diffraction from specific precipitates, or corresponding selected-area EDS analysis shown in the inserts of Fig. 9(d) and (e). Finally, it should be mentioned that the extent of transformation implicit in the LM views of sensitization attack can also be examined by X-ray diffraction which provides a "global" view of the associated microstructures, and these features are illustrated in Fig. 9(f).

Summary and Conclusions

We have presented four case examples to illustrate some of the important aspects of multi-dimensional microanalysis, and the necessity to examine microstructural issues utilizing different techniques which provide different views, even spatially dimensional views. In the examples described, it is readily apparent that TEM and associated SAD techniques offer possibly the most powerful analytical and microanalytical approach to materials characterization. This conclusion is to some extent aphoristic since the same conclusion is now legion in the materials science and engineering community, and has been self-sustaining and self-reinforcing for decades [1,2]. As obvious as this may appear, this aphorism is not necessarily pervasive, and there continue to be numerous examples of ambiguous if not inaccurate conclusions drawn about materials structures and microstructures because only a single mode of analysis is applied. This presentation is but another appeal to more effective materials characterization through comprehensive and multi-dimensional approaches, especially involving TEM analysis.

Acknowledgements

We are grateful to Dr. M. A. Meyers for providing the tantalum shock loaded specimen examined in a portion of one of the case examples. The research represented in the examples described has been supported in part by a Mr. & Mrs. MacIntosh Murchison Endowed Chair, NASA Johnson Space Center Grant NAG-9-481, GSA Contract PF90-018, U.S. Army Contract DAAA21-94-C-0059, and the Patricia Roberts Harris Fellowship Program (R.J.R. and J.G.M.)

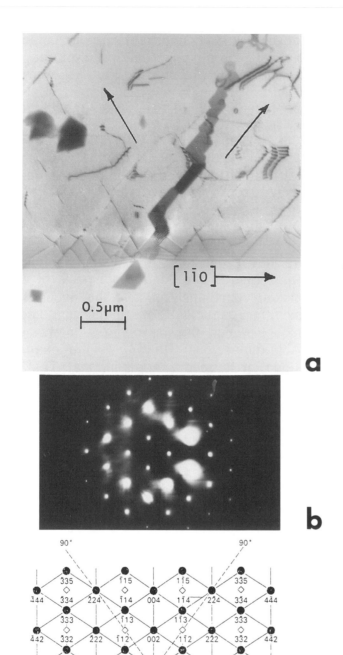

Fig. 8 - TEM bright-field image of lamellar carbide ($M_{23}C_6$) growth in 304 stainless steel (10h at 750°C) (a). The SAD pattern (b) shows a (110) grain surface orientation characterized in the diffraction net in (c). The dotted lines show the trace of {111} planes at 35° and 90° respectively. Note that the precipitate grows preferentially in the <112> directions from the (111) twin plane along [1$\bar{1}$0].

References

1 Murr, L. E., "Electron Optical Applications in Materials Science", McGraw-Hill, New York (1970).

2 Murr, L. E. "Electron and Ion Microscopy and Microanalysis: Principles and Applications", 2nd Edition, Marcel Dekker, Inc., New York (1991).

3 Wang, S-L., and L. E. Murr, Metallography, 13, 203-224 (1980).

4 Anderson, R. W. and S. E. Bronisz, Acta Met., 7 645-647 (1959).

5 Barrett, C. S. and R. Bakish, Trans. AIME, 212, 122-125 (1958).

6 G. E. Dieter, "Rsponse of Metals to High Velocity Deformation", Wiley Interscience, New York, 1961.

7 Wittman, C. L., R. K. Garrett, Jr., J. B. Clark, and C. M. LoPatin, "Shock Wave and High-Strain-Rate Phenomena in Materials, p. 925, M. A. Meyers, L. E. Murr, and K. P. Staudhammer, eds., Marcel Dekker, Inc., New York (1992).

8 Wongwiwat, K. and L. E, Murr, Mater. Sci. Engr., 35, 273-285 (1978).

9 Murr, L. E. and W. Kinard, American Scientist, 81, 152-165 (1993).

10 Quinones, S. A., J. M. Rivas, and L. E. Murr, "Microstructural Science", ASM International, Materials Park, Ohio (1995).

11 Rivas, J. M., S. A. Quinones, and L. E. Murr, Scripta Metall. et Materialia, in press (1995).

12 Murr, L. E. and J. M. Rivas, Int. J. Impact Engrg., 15(6), 785-795 (1994).

13 Bernhard, R. P. and F. Horz, Int. J. Impact Engrg. in press (1995).

14 Rivas, J. M., L. E. Murr, C-S. Niou, A. H. Advani, and D. J. Manuel, Scripta Metall. et Materialia, 27, 919-924 (1992).

15 Murr, L. E., "Shock Wave and High-Strain Rate Phenomena in Metals", p. 570, M. A. Meyers and L. E. Murr, eds., Plenum Publishing Corp., New York (1981).

16 Meyers, M. A., "Dynamic Behavior of Materials", Wiley, New York, (1995).

17 Romero, R. J. and L. E. Murr, Acta Metallurgica et Materialia, 43(2), 461-469 (1995).

18 Murr, L. E., R. J. Horylev and W. N. Lin, Phil. Mag. 20, 1245-1258 (1969).

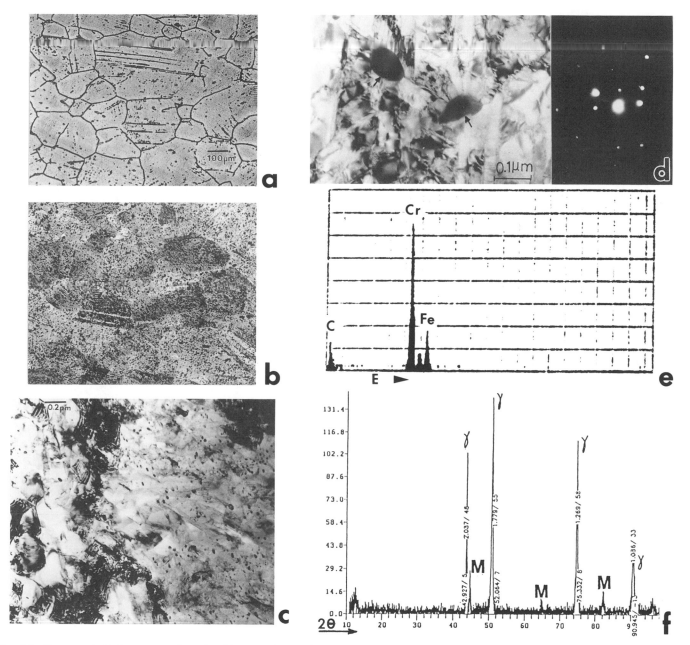

Fig. 9 - $M_{23}C_6$ precipitation associated with transformation and sensitization in 304 stainless steel. (a) Degree-of-sensitization test attack image (LM) of precipitation along grain and twin-related boundaries after uniaxial straining 20% at room temperature (300K) and aging 1h at 670°C. (b) Test attack image for sample uniaxially strained 20% at 78K and aged 0.1h at 670°C. (c) TEM image for (b) showing carbides at 2-phase interfaces. (d) TEM image of carbides and microdiffraction pattern $\overline{2}3\overline{7}$. (e) EDS spectrum typical for precipitates in (d). (f) XRD spectrum for sample as in (b) but aged 0.4h at 670°C. M indicates bcc martensite.

19 Advani, A. H., L. E. Murr, D. J. Matlock, W. W. Fisher, P. Tarin, C. Ramos, R. J. Romero, R. L. Miller, J. G. Maldonado, and C. M. Cedillo, Scripta Metall. et Materialia, 28, 1155-1160 (1993).

20 Murr, L. E., Interfacial Phenomena in Metals and Alloys, Addison-Wesley Publishing Co., Reading, Mass. (1975); reprinted by Tech Books, Herndon, VA (1990).

Reflections on Incident Light Microscopy

James H. Richardson
The Aerospace Corp., Los Angeles, CA

Abstract

The metallograph, specifically an incident-light microscope, is a venerable tool in the investigation of metallurgical structure. Newer technologies using X-rays, electrons, and surface energy may seem to have overshadowed the metallograph in importance. However, this paper will show that the metallograph continues to evolve and is still a major tool.

Introduction

THE INCIDENT-LIGHT MICROSCOPE became a significant instrument for metallurgical examination in the hands of Henry Clifton Sorby (1). He was one of the first in a long heritage of researchers, engineers and technicians using the incident-light microscope, usually referred to as a metallograph, to bring the field of metallurgy to the state we see today. This is not to diminish the importance of information derived from the newer electron, X-ray and surface-sensitive instruments, but the metallograph usually provides the first insight about an alloy or material and has served as the primary diagnostic tool in metallurgy for many years.

It may seem to the casual observer that little has changed with the light microscope over the years. In my two books on microscopy (2,3), I have had the opportunity to chronicle the advances in both optical and mechanical design of the metallograph. At the completion of each of these efforts, I would briefly share the feelings of Charles Duell, a Commissioner of the U.S. Patent Office, who in 1899 stated that, "Everything that can be invented has been invented." (4) More recently, Hornbogen and Petzow (5) have stated,

"Although optical metallography is becoming increasingly widespread it cannot boast any notable innovations." They do acknowledge incremental change and it is that change which is the basis for this present review. My preparation for this paper indicates that once again there are some new aspects to metallographs!

Advances, and changes, are occurring in the microscope industry. During my time as a microscopist we have seen the Japanese instruments rise to world importance; we have seen the rejoining of split giants like Zeiss; the effective demise of old name organizations such as B&L, American Optical, Wild, Reichert, and Vickers; and merger upon merger to give us organizations such as Leica, Inc.

Not so long ago, the crown jewel of any metallographic laboratory was a research metallograph, an instrument that consumed and was, in fact, an integral part of a full bench. Two of these instruments that really impressed me were the B&L Research metallograph and the Leitz MM-5 metallograph; the latter was particularly impressive to me since I had the opportunity to use it. Aside from the obvious elegance and prestige that such instruments lent to the laboratory, they were usually equipped with all the accessories necessary for various types of illumination, and were extremely well built so that they did not succumb to every vibration that shook the laboratory. It is sad to note the passing of these research instruments; the last of these, the Neophot 32, manufactured by Zeiss Jena is no longer available. These grand instruments have been supplanted by smaller bench type instruments.

Typical of the bench type instruments currently in vogue are two Zeiss instruments; an upright instrument and an inverted instrument, shown in Figures 1 and 2.

The inverted instrument is generally the choice of the metallographer because no sample leveling is required.

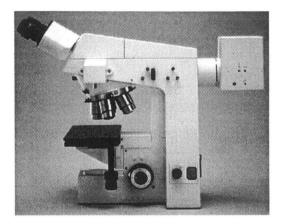

Fig. 1. The Zeiss Axioskop is a typical example of an upright metallographic microscope that is capable of bright-field, dark-field, and differential interference contrast (DIC) illumination. Courtesy of Carl Zeiss, Inc.

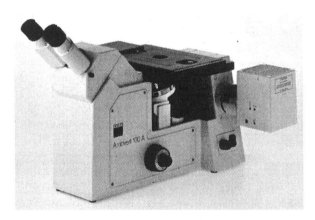

Fig. 2. The Zeiss Axiovert 100A is a typical example of an inverted metallographic instrument that is capable of bright-field, dark-field, DIC, polarization, and reflected light fluorescence illumination. Courtesy of Carl Zeiss, Inc.

The upright microscope is preferred by those in the semiconductor industry because their samples generally do not require leveling and the sample surfaces of interest may be damaged by placing them on the stage of an inverted microscope.

The great technological marvels to be chronicled in the remainder of the paper were not achieved without considerable investment on the part of the purchaser. Figure 3 is a woodcut of a fine instrument offered in 1884 by W. H. Walmsley & Co., sole U.S. distributor for R. & J. Beck of London. This instrument, the "International" Improved Large Best Microscope, came completely equipped with all of the latest accessories for the lordly price of $1650, whereas Figure 4 shows a modern state of the art, fully equipped research microscope which costs about $95,000. Assuming a

modest inflation of 3.7%, these microscopes are roughly equivalent in price, but not in capability.

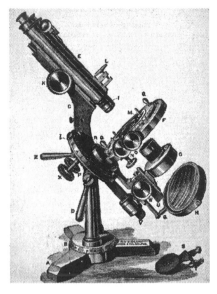

Fig. 3. A woodcut of the "International" Improved Large Best Microscope.

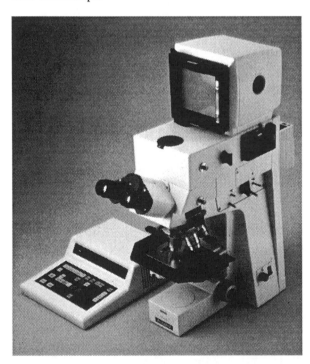

Fig. 4. The Zeiss Axiophot Photomicroscope is typical of a modern metallograph, incorporating all common types of illumination. This instrument has ports for three separate film cameras, a port for video camera, and simultaneous binocular viewing. Courtesy of Carl Zeiss, Inc.

While we are thinking about innovation and change we must not overlook the effect of the computer. Firmly woven in the fabric of this discussion of advances in microscopes is the role of the computer.

Witness the alphabet soup: computer aided design (CAD) is used in defining the lenses and stands; computer aided manufacturing (CAM) finds application for improving the quality and reducing the cost of instruments; Computer Aided Enhancement (CAE) provides improvement of the microscope images; and quantitative image analysis (QIA) allows for easy and detailed quantification of a metallographic structure.

At the conclusion of this report I hope not to fall prey to the "Duell Syndrome," and will expect to see a sequel to this paper in perhaps a decade documenting newer and more improved metallographs.

The Metallograph

The Objective Lens. It is well understood that the objective lens is that part of the microscope which ultimately defines the instrument's quality, and as a result the manufacturers routinely expend their greatest effort in producing high-quality objectives.

The refractive objective lens consists of a threaded mount containing two or more glass lens elements, each with specific refractive index and dispersion properties. The lenses must be of the proper diameter, with correct curvatures and critically controlled spacings. These lenses are coated with antireflection surfaces, very often of different composition on each side of the lens. The individual lenses may be as small as 0.5 mm in diameter; a size that is a challenge both for fabrication and coating of the lens as well as in its sturdy placement within the mount. A typical cross section of a high-quality lens is shown in Figure 5.

Fig. 5. Cross-section of a typical high-quality objective lens showing the various glass elements. Courtesy of Carl Zeiss, Inc.

We, as microscopists, expect a great deal out of our objective lenses with little or no thought to their construction. Our list of requirements goes on and on: we want a sharp image, high contrast, high resolution, flat field, long working distance, no distortion, and of course a low price. To achieve this is no small task. Added to the more obvious factors above, the manufacturer must consider spherical and chromatic aberration, off-axis aberrations, the number of glass elements that he wants to use and the availability of glasses of suitable optical properties. In years past, the

complexity of the hand calculations of the multiple lens formulas limited what might be expected. Months could have been involved in the ray tracing for a single lens. Even then, you would be limited; you could get an objective lens that had apochromatic or superior color correction and high resolution, but with a field of view that was noticeably curved. Otherwise you could get an objective lens with a fairly flat field, but with little color correction.

The computer is now able to reduce the time for complicated lens calculations that took months in an earlier time, to days with the complex lens design programs now available. Furthermore, the equations may be solved for the best solution for a greater number of parameters. These programs permit looking at "what-ifs"; i.e., looking at a large number of different variables, with results that are truly astonishing. Today, we have a large selection of really good objective lenses available. For example, it is possible to have a metallographic objective that has both apochromatic correction and a flat field.

Another feature of more and more objective lens systems is infinity correction. With these objective lenses the rear focal plane is placed at infinity; in other words, the rays leaving the objective are parallel. Thus, any parallel-plate accessories inserted in the image path will not introduce aberrations into the final image. After passing through these accessories, the image rays are then brought to a usable focus by a tube lens at a point before the eyepieces.

Remaining Optics. After the image is formed by the objective lens, its rays will pass through the beam splitter that provides an entrance for the illumination. From there, one may find various other accessories, such as accessory plates, polarizers, barrier filters, and possibly a tube lens if the objectives are of the infinity corrected type. The image rays then proceed through the eyepiece(s) to the observer's eyes or a camera.

The Micromate, a new instrument from Navitar, is interesting in that it has no eyepieces and its only output is video; thus the image can be displayed on any suitably sized monitor. This instrument is designed to use infinity-corrected objectives and can function as a macroscope or a microscope with magnifications ranging from 14 to over 4000X on a 13-inch monitor. A built-in illumination source provides for normal incident, oblique incident and transmitted illumination. This microscope, shown in Figure 6, would be a superior instrument to reduce eye fatigue in routine metallographic examinations, such as for plating-thickness measurements or evaluation of electronic printed-circuit-board mounts. With its normal video output, this microscope is also suitable for use with image enhancement, image analysis, and video recording accessories. The Micromate can serve an important function by providing real-time examination of samples during video teleconferencing with a single compact instrument.

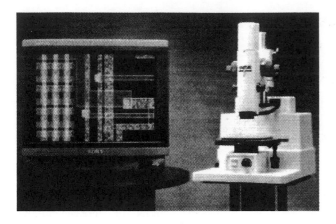

Fig. 6. Micromate - This microscope/macroscope has no visual capability; the only output is video. This instrument should be very useful for inspection. Courtesy of Navitar, Inc.

Fig. 7 Arrangement for a typical image analyzer. Courtesy of Buehler, Inc.

The Stand. As noted above, the research bench instruments are no longer available and have been replaced with smaller bench type instruments. These stands are well designed with ergonomics in mind. Only a scant few of these require you to rest your chin on the photographic camera; this requirement is a particular annoyance to me. Some of these microscopes are also equipped with armrests that are very comfortable for protracted examinations. Perhaps one of the more innovative features of recent times is Nikon's variable-inclination binocular body featured on their Optiphot-2 and their Labophot-2 microscopes. With a viewing angle that is adjustable from 10 to 40 degrees, back and neck strain should be reduced to a minimum if the microscope is properly positioned and a correct seat height is used.

Accessories. The 1960s and 1970s were probably the apex of microscope accessory availability. In my first book, a large number of accessories were listed that attached to various parts of the microscope. High- and low-temperature stages, particle counters, cameras, etc., were available from either the original equipment manufacturer or from aftermarket sources. The number was much less in my most recent book, and I believe that there are even less today, if the advertisements in periodicals such as *The Microscope Book, Microscopy and Analysis, the RMS Proceedings* and of course, *Materials Characterization* are any indication.

Probably, the accessories of greatest interest today, at least for the metallographer, are the image analysis systems, such as shown in Figure 7. The major manufacturers of microscopes often have their versions; however, a number of excellent after-market accessories, such as the Buehler Omnimet are available.

The automatic point counter, model G, manufactured by Prior Scientific and James Swift is an improved computer-controlled version of one of the very early instruments for quantitative analysis of structures. This instrument uses a stepping stage for accurately translating the sample, and relies on the observer's eye for phase identification and discrimination. This instrument can still prove useful in those cases where differences in structure are simply too subtle for electronic resolution. One feature of this update version is a provision for transferring data from the model G to a host computer for further processing.

Applications of image analysis systems are discussed elsewhere in this symposium, so I will not dwell on them, other than to emphasize that their capability and flexibility depend on the use of ancillary computer systems.

Prior Scientific Inc., Ludl Electronic Products Ltd., and New England Affiliated Technologies, as well as others, manufacture a number of programmable, stepper-motorized X-Y stages suitable for metallurgical applications. Various ranges of travel and step repeatability are available for these stages. A typical stepper system is shown in Figure 8. Among the other motorized devices available are a rotating table manufactured by Prior Scientific and an autofocus system developed by Ludl.

Contrast Enhancement

Traditional Contrasting Techniques. The best bright-field microscope with the highest resolution optics will not give usable information about a sample that exhibits no contrast. It is the old story of trying to see "the polar bear eating marshmallows in a snow storm." This section will chronicle the methods by which the contrast of a sample can be enhanced optically.

Contrast with a bright-field microscope has been traditionally achieved by building it into the sample; a biological sample may be differentially stained, or

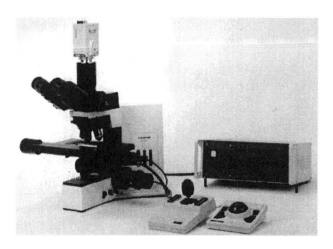

Fig. 8. Microscope equipped with a stepper stage for rapid positioning of the sample on the stage of the microscope. Courtesy of Prior Scientific, Inc.

polished metal sample may be etched. In the case of metals, a grain boundary etch does not appreciably affect the polish, while the grain boundaries are attacked preferentially because they have a higher density of impurities or dislocations. The result as observed under the metallograph is that each bright grain is delineated with a black boundary. Other etches may affect grains of differing orientation differently; some grains may be surface etched, affecting their reflectivity, or they may be etched in different colors. In all cases each grain should be clearly distinguished.

In those cases where etching is poor or ineffective, the metallographer has resorted to microscopical techniques to achieve contrast. The early techniques of oblique and dark field illumination are still used but they have limitations. Oblique illumination is less than ideal because of its azimuthal variation of contrast and dark field illumination has low light levels for photomicrography. Phase contrast is sensitive and has satisfactory light levels. Incident light phase accessories are expensive and difficult to use, therefore they are seldom used today and no longer offered for new instruments. Cross-polarized illumination may be used effectively for bireflecting materials. An interesting new application for polarized light is being offered in some of Zeiss' instruments; namely, circular polarization (6). This technique has found most application for translucent materials as both an analytical tool and as a contrasting method. I believe that circular polarization can be more useful as a contrasting method for incident-light applications than cross-polarized illumination because no grains in the image will appear fully extinguished or dark.

The interference-contrast accessory is the current favorite for building contrast into an image. The interference contrast method, like the phase contrast method, relies on very slight differences in the relative heights or elevations of the etched grains on the sample surface; these slight differences in height would be insufficient to produce any visible effects in bright field illumination. By appropriate adjustment of the interference-contrast accessory, the image may be varied from what appears to be a dark field image, to one that is similar to a phase contrast image, to one where the grains are contrasted in any one of a number of color combinations. While the contrasting capability of the interference-contrast technique is outstanding, one must be careful in the interpretation of the image for anything beyond delineation of grain boundaries.

Electronic Contrasting Techniques. The traditional optical contrasting techniques generally operate on aspects of the image information which are not visible to the eye, such as the phase and polarization components of the rays reflected from the sample. The newer electronic enhancement techniques amplify or enhance differences from area to area that tend to be too subtle for the eye to see. In this type of contrasting, a photomicrograph may be operated on to produce the enhanced image; however, in this case the photographic process may limit the amount of increased contrasting available. The greatest flexibility can be realized by operating directly on the output of a video camera attached to the metallograph.

Microscope manufacturers have developed turnkey electronic contrasting systems. For example, Leica has offered their Multicon system which included a black-and-white video camera, their Multicon instrument and a RGB color monitor. Electronic contrasting systems are also available from after-market manufacturers for addition to existing metallographs. For example, Semicaps has a complete digital imaging system that can be interfaced with a metallograph, and all types of electron and X-ray imaging systems. Some of their capabilities for metallography are enhancement, analysis, and storage of the image as well as the capability for transmission of images or collections of images via phone, E-mail and network. Their image processing includes brightness and contrast correction, gamma, smoothing, sharpening, pseudocolor, addition, subtraction, zoom, aligning and annotation of the image with text and arrows. Needless to say, these various operations rely on a computer-based program.

It is worthwhile to suggest a user-constructed, electronic contrasting system which would be capable of good to excellent output; a typical layout for such a system is shown in Figure 9. The microscope image is viewed with a CCD video camera located in the traditional camera port; cameras with various resolutions are available. The output of that camera is accepted by a frame capture board in a nearby computer. This computer then can be used to manipulate the image as indicated above. Examples of suitable programs for this manipulation are Corel PHOTO-PAINT for the PC and Adobe Photoshop for both the Macintosh and the PC. The file thus created can be exported to a drawing or page layout program on the same computer, where callouts, text, titles, etc., can be added and multiple images can be collected on a single page. The resulting file can then be printed. The printer can be a black-and-white laser printer or a dye-sublimation printer for color prints. The file can also be sent to a commercial prepress shop for higher resolution color printing. The cost of such a system will vary greatly based on the desired output quality and whether printing is done in-house or sent out. A sample of images prepared using this approach are shown in Figure 10.

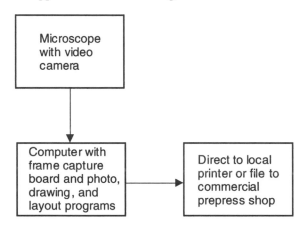

Fig. 9. Instrumentation and software for electronic image enhancement.

Confocal Microscopy. Confocal microscopy, described in detail in the literature (7,8), is commonly the domain of the biological microscopist. Excellent results have been obtained from delicate samples, thick sections, and even living tissue.

Confocal microscopy can be achieved in several different ways; but in the most recent version, a spirally perforated disc, known as a Nipkow disc, is rotated in the illumination path in a plane that is conjugate with the plane of focus in the sample. Thus, only those areas of the sample in the exact plane of focus are illuminated. By changing the sample's position with respect to the plane of focus—i.e., focusing through the sample—the eye or camera sums the individual images and integrates the information into an image that displays the three-dimensional character of the subject.

The confocal technique can also be used in conjunction with most of the contrasting techniques discussed above and for incident-light microscopy, as well. Entwhistle (9) has demonstrated that confocal microscopy can be used to study voids in polished ceramic samples. I feel sure that this technique could be extended to study solid second phases and even negative crystals in translucent materials (10).

Image Recording

Photographic Imaging. There is little to indicate that much has taken place recently in the field of silver imaging, however, it is still the medium of choice for critical research and publication. An inspection of

manufacturer's literature will show the fine touch of computer aided manufacturing (CAM) to refine the grain

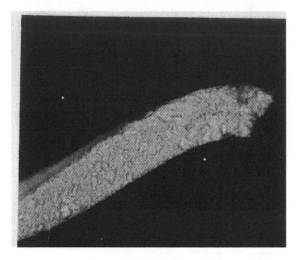

(a) Photomicrograph of a silver ribbon with normal contrast and sharpness.

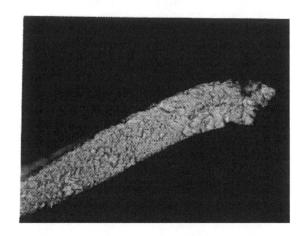

(b) Same area with increased contrast.

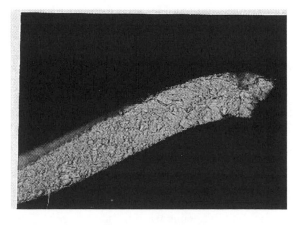

(c) Same area with increased sharpness.

Fig. 10. A series of photomicrographs showing the various effects of electronic contrasting. Courtesy of Semicaps, Inc.

size of the light-sensitive silver halide grains, thereby giving greater resolution and more uniform exposure over the photomicrograph.

In the area of instant photomicrography, the dominant force is still Polaroid. They have expanded the size of their prints to 8" x 10" and now provide the MicroCam, a more advanced single-lens reflex camera that mounts onto the eyepiece of a microscope. The exposure is controlled by a small on-board computer. Polaroid has marketed at least three generations of these devices.

Polaroid prints have improved over the years in resolution. Contrast is better for both the black-and-white format and for color prints. Their color rendition has improved markedly. As a result, Polaroid prints have become very acceptable for routine work

Electronic Imaging. Computer printers, not so long ago, were of the daisy wheel and the dot-matrix types that for all of their noise, produced only a mimic of the typewritten page, or worse. The bit-mapped printers with their higher resolution brought in the era of desktop publishing and the capability of adding graphics to printed documents. The early line drawings quickly gave way to images with significant gray scales and then color. The coupling of video imaging instrumentation with these printers was inevitable; it was then only a brief step to recording of micrographs using this electronic format. To be sure, the resolution and contrast of these prints are not good by photographic standards, but the advantages of this medium may cause us to overlook the disadvantages. For example, the same computer that grabbed the image from the video camera could provide manipulation of its size, contrast, gray scale, and color rendition. Furthermore, with very little additional effort, several of these electronic pictures can be combined with captions and callouts for inclusion in a document and, if speed dictates, can be sent on their way by E-mail. A typical output of such a system is shown in Figure 11.

Much of the electronic capability for image enhancement has also been integrated into the image analyzer as mentioned above. In this way, hours of eye-fatiguing and brain-numbing work with an integrating stage or an integrating eyepiece has been reduced to minutes. This is another of the revolutionizing combinations of the microscope with the computer.

Summary

From the preceding, it is clear that there is no regression with respect to metallographs, with perhaps the possible exception of the physical size of the instruments. On the positive side, there are a number of aspects of the metallograph in which we can see measurable gain. The stands, while generally smaller are more functional and ergonomically correct. Even though we are near the limit in resolution, overall

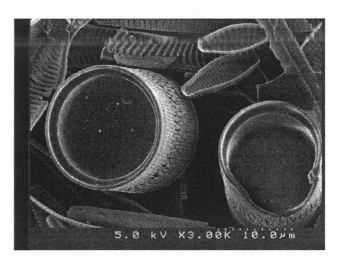

5.0 kV X3.00K 10.0μm

Fig. 11. An example of electronic image enhancement using a standard image-scanning system and computer.

quality of objective lenses is steadily improving. Electronic imaging, recording, enhancement, transmission, and analysis are easier and more affordable. Thus, we can see the sample better and can record the images in new ways. With the aid of the computer we can combine electronic images on a page and provide "on page" captions, as well as, callouts on the pictures themselves and transmit them electronically around the world, if necessary.

References

1. Smith, C. S., "A History Of Metallography," p.169, Univ. of Chicago Press, (1960).
2. Richardson, J. H., "Optical Microscopy for the Materials Sciences," Marcel Dekker, Inc., NY (1971).
3. Richardson, J. H., "Handbook for the Light Microscope, A User's Guide," Noyes Publications, Park Ridge, NJ (1991).
4. Cerf, C., "The Experts Speak: Definitive Compendium of Authoritative Misinformation," p. 203, Pantheon Books (1984).
5. Hornbogen, E. and G. Petzow, Prakt. Metal., 28, 383 (1991).
6. Zeiss, C., Micro Info, Edition 33, Jun (1993).
7. Sheppard, C. and D. M., Shotton, "Confocal Laser Scanning Microscopy," Royal Microscopical Society Handbooks, BIOS Scientific Publishers, (1995).
8. Wilson, T., Editor, "Confocal Microscopy," Academic Press (1990).
9. Entwhistle, A., Proceedings of the RMS, 29/3, Jun (1994).
10. Richardson, J. H., Negative Crystals in Beryllia (A Progress Report). Presented at the Gordon Research Conference on Solid State Studies in Ceramics, Jul 31, 1960.

Progress in Characterizing Materials Using Image Analysis

George F. Vander Voort
Carpenter Technology Corp., Reading, PA

Abstract

During the past ten years there has been an exponential growth in the power of personal computers and work station computers as well as vast improvements in high resolution cameras and monitors. Image analyzers have become much more powerful, much faster, fully automated and more user friendly with software customized for many applications. Stereology and metrology matured and mathematical morphological procedures are being utilized more commonly, particularly for image segmentation. This capability has become integrated into research and quality control laboratories and is being utilized to characterize metals and materials. Such usage has stimulated the development of standardized test methods and demonstrations of the precision and reproducibility of measurements and detection of measurement bias. ASTM Committee E-4 on Metallography has developed a number of standard image analysis-based test methods. Standard chart rating methods, such as ASTM E 45 for inclusion rating, have been more formally defined and revised so that image analysis can be employed (ASTM E 1122). A stereological approach for characterizing any discrete second phase has been developed, ASTM E 1245. An image analysis grain size standard, E 1382, has also been developed. These three test methods are discussed.

SINCE THE INTRODUCTION of television-based image analysis technology in 1963, there has been a sustained growth in the technology which accelerated dramatically since the introduction of PC technology about ten years ago. The incredible growth in computer power transformed top-of-the-line image analyzers from hardware-based systems to hybrid software-based systems using high speed processors on frame grabbers to totally software-based systems.

Complementing the growth in computer power was a similar growth in video technology - increased resolution cameras and monitors, advances in solid-state camera technology (chip size, pixel density, light sensitivity), and faster, higher-pixel density frame grabbers. All of these ingredients are being combined and controlled with more user-friendly, more flexible, more capable software programs.

While the above advances are relatively well known to the metallurgical community, and to those in many other scientific disciplines, little has been published about standardized procedures for using image analyzers to provide reliable data. A modern image analyzer, coupled to a high quality research type light microscope with an automated stage and automatic focus, can produce an incredible amount of data from a single metallographic specimen. Indeed, one can be overwhelmed with data.

With all this power at our command it is important to establish universally accepted routines for the systematic measurement and analysis of microstructures. Chaos would result if everyone developed and used their own measurement and analysis schemes. To a large extend, the application of image analysis to the measurement of microstructures is controlled by existing materials testing standards, many of which evolved from manual chart methods, for example, ASTM A 247 (graphite rating) and ASTM E 45 (inclusion ratings). Others are combinations of chart methods and manual stereological measurements, such as ASTM E 112 (grain size). Only a few are based solely on the principles of stereology, such as ASTM E 562 (point count method for determining volume fractions of phases).

ASTM Committee E-4 on Metallography has had a

Table 1. Original Limits for ASTM E 45.

Severity	A (μm at 1X)	B (μm at 1X)	C (μm at 1X)	D (Number)
0.5	38.1	38.1	38.1	1
1.0	127.0	76.2	76.2	3
1.5	254.0	177.8	177.8	9
2.0	431.8	304.8	304.8	14
2.5	635.0	508.0	508.0	20
3.0	889.0	812.8	762.0	26
3.5	1143.0	1168.4	1016.0	35
4.0	1524.0	1524.0	1270.0	44
4.5	1905.0	2032.0	1778.0	52
5.0	2286.0	2540.0	2159.0	64

long history [1,2] of involvement in developing standardized test methods for measuring microstructural features. Although subcommittee E.04.14 on Quantitative Metallography was formed in 1960, the first measurement procedure was published in 1917 when ASTM E 2 was published, only a year after E-4 was organized. There are now twelve ASTM standards involving microstructural measurements under E-4 jurisdiction. Several of these involve image analysis and three are discussed in this paper.

Inclusion Measurement

Because of the influence of nonmetallic inclusions on the properties of metals, which has been thoroughly studied for steels, the measurement of inclusions in engineering alloys used in critical applications has been common practice for more than sixty years. While a vast number of measurement procedures have been proposed and used over the years [3], chart methods have become the dominant procedure and the JK (Jernkontoret) chart, developed in Sweden [4], is widely used. When ASTM E 45 [5] was developed in 1942, it incorporated the JK chart which illustrates five degrees of inclusion content and two series of thicknesses, or diameters, for four basic types of inclusions. In 1963, ASTM E-4 developed a modification of this chart for steels with lower inclusion contents, such as vacuum degassed bearing steels. These charts are also used in ISO 4967 [6] and in national standards developed by a number of other countries.

Over the years, these methods have gained popularity and widespread usage. However, they are not without

their problems and their use has garnered substantial criticism [7]. Charts are used in two basic ways. First, we can obtain picture ratings of the worst condition of each inclusion type and thickness present. Second, we can rate the inclusions in every microscopical field within a certain test area size. The first method is rather insensitive while the second method produces a great deal of data, difficult to summarize and utilize, and rather time consuming to obtain. Chart ratings have been shown to lack repeatability and reproducibility in numerous round-robin programs. Classification errors are also a problem and the images do not always relate to inclusions in today's steels.

To counter these problems, Committee E-4 has been working on procedures to improve JK ratings by using image analysis. The feasibility of performing JK ratings by image analysis was demonstrated in 1981 [8]. The method was developed as ASTM E 1122 [9], first issued in 1986. This standard has been updated in 1992 and 1995 as image analysis technology has improved. Several manufacturer's of image analysis equipment offer custom-made software for performing E 45/E 1122 inclusion ratings. Modern systems can perform a complete quantitative (method D) rating in only a few minutes. Image analysis has been shown to yield reproducible, repeatable measurements [10]. Most of the problems associated with manual JK ratings are overcome by the use of image analysis.

To improve the sensitivity of JK ratings, especially for steels with low inclusion contents, it was necessary to modify the inclusion measurement - severity level relationships so that the inherent value of actual measurements by image analysis could be exploited. The original measurement limits, Table 1, were plotted

Table 2. New Limits for ASTM E 45 and E 1122.

Severity	A (μm at 1X)	B (μm at 1X)	C (μm at 1X)	D (Number)
0.5	37.0	17.2	17.8	1
1.0	127.0	76.8	75.6	4
1.5	261.0	184.2	176.0	9
2.0	436.1	342.7	320.5	16
2.5	649.0	554.7	510.3	25
3.0	898.0	822.2	746.1	36
3.5	1181.0	1147.0	1029.0	49
4.0	1498.0	1530.0	1359.0	64
4.5	1898.0	1973.0	1737.0	81
5.0	2230.0	2476.0	2163.0	100

on log-log paper and a straight line was fit to the data by the least squares method. The fit was quite good for the A type inclusions, but somewhat poorer for the B, C and D types.

The A, B and C types require a measurement of inclusion or stringer length per field to define the severity while the D type requires a count of the number per field. For the A, B and C types, we could simply move the data points slightly to obtain a perfect log-log correlation between the length measurement and the severity.

This, of course, was not possible for the D types where the severity varies with the count per field, i.e., discrete numbers. Because the most important part of the count-severity relationship, from the standpoint of specifications, is in the 1 to 2 severity range, we wanted to keep these numbers as similar as possible. The original test established that the severity was 0 when there were no rateable (globular inclusion ≥ 2 μm diam.) inclusions in a field and was ½ when there was one rateable inclusion in the field. Consequently, as these could not be changed, the new relationship for severities of 0, ½, 1, 1½, ... 4, 4½, 5 was established based on the simple power series of 0^2, 1^2, 2^2, 3^2,... 8^2, 9^2, 10^2, that is 0, 1, 4, 9, ..., 64, 81, 100 inclusions per field, Table 2.

This resulted in no change for severities of 0, ½ and 1½, minor changes for severities of 1 and 2, and larger changes as the severity increased above 2. Although the most critical range from a severity of ½ to 2 changed very little, many people expressed considerable concern about these small changes. However, experience with round robins with manually generated JK ratings clearly demonstrated that these minor limit changes would never be noticed. The new D series does permit definition of 101 unique severity values as the number of inclusions

per field increases from 0 to 100.

For the A, B and C types, severity levels can be defined in 0.1 increments from 0 to 5, rather than in ½ or 1.0 increments as done manually using either Plates I or III of E 45. This, of course, cannot be achieved for D types. Note that no severity values can exist between 0 (no inclusions) and ½ (1 inclusion); but, a distinct severity value exists for any number of inclusions per field between 0 and 100.

Because the image analyzer separates the inclusions in each field into the types by gray level and shape differences using the same procedure each time and severities are based on actual measurements or counts, results are highly reproducible and repeatable. Additionally, modern image analyzers can operate with minimal operator interaction [10]. As a result the E 1122 approach for rating inclusions has become very popular.

Measurement of Second-Phase Particles

The characterization of discrete second-phase particles, such as inclusions, carbides, nitrites, borides, sigma, etc., can be performed using manual stereological methods, such as point counting, to obtain their volume fraction. While procedures for determining the volume fraction have been standardized (see, for example, ASTM E 562 or ISO 9042), determination of other useful stereological parameters has not been standardized. Many people have proposed using one or more stereological (and non-stereological) procedures for characterizing inclusions [3] and several early attempts were made (unsuccessfully) to standardize these approaches.

Researchers have realized that inclusions could be better characterized by stereological measurements but manual implementation of even the simplest was impractical. For example, the writer once evaluated the volume fraction of inclusions in a series of specimens with a wide range of sulfur contents using manual point counting and lineal analysis [3]. After one hour of work per specimen, the relative accuracies of the measurements were barely acceptable, but only for the highest volume fractions. As the inclusion (or second phase) content decreases below 2%, the amount of effort required to obtain a 30% relative accuracy (about as good as one can expect) increases, and may not be achievable below a certain inclusion content. Clearly, this is a problem where image analysis can be most useful, especially if the stage is automated (in X, Y and Z).

To characterize discrete second-phase particles, we need more than just a volume fraction. It would be useful to have a particle count (particle "density"), but this must be related to a specific measurement area (for example, as the number per mm^2, N_A, of test area). Some idea of the dimensions of the particles (on the cross section through the particles, the plane of polish) is required. The average area and a diameter or length are very useful. Finally, an idea of the spacing of the particles is quite helpful. The "mean free path" (λ), the mean edge-to-edge spacing between particles, has been found to be a useful structure-sensitive parameter.

To develop a true estimate of the three-dimensional nature of the particles, we need to make these measurements on the three principle test planes; that is, the longitudinal, transverse and planar surfaces of a plate-like shape. However, if the particles are equiaxed in shape, measurements on any plane of polish will yield statistically similar results. If the true three-dimensional aspects of the particles is not required, all measurements can be performed on the same plane. The longitudinal plane is most commonly chosen as it will reveal any departure from an equiaxed shape. As long as all measurements are performed on the same plane, results can be compared.

This is the approach taken in ASTM E 1245 [11]. Measurements are made on the longitudinal plane to determine the area fraction, A_A (which is equivalent to the volume fraction, V_V) in %, the number per unit area, N_A, in No./mm^2, the average area, A, in μm^2, the average length, L, in μm, and the mean free path, λ, in μm. In this work, the length is in the direction parallel to the deformation axis and the mean free path requires a determining of the number of particles intercepted by test lines perpendicular to the deformation axis, N_L. Thus, λ is the mean edge-to-edge spacing between particles in the through-thickness direction. The mean center-to-center particle spacing could also be determined but it does not account for the particle size and is considered to be not as sensitive to properties as λ, the mean edge-

to-edge spacing. λ is calculated according to:

$$\lambda = \frac{1 - V_V}{N_L} \quad (1)$$

where the volume fraction is a fraction rather than a percent, N_L is the number of interceptions per mm of test line length, and λ is expressed in μm.

The determination of the mean free path does present some problems if the amount of the constituent is quite low and blank fields can be encountered. In this case, λ must be calculated after n number of fields are measured, that is, at the end of the run. It cannot be calculated for each field if no inclusions are present on certain fields. Thus, for a given specimen, a mean value of λ can be obtained but the standard deviation cannot be determined. The mean value and standard deviation can be determined after n measurements of V_V and N_L, but only for these parameters. However, if the constituent is present in every field, then λ can be calculated for each field and the mean and standard deviation of λ can be computed from the individual field values.

If individual particle measurements are made of the area and length of each particle, the mean and standard deviation of A and L can be determined based on measurement of N particles in n fields. The mean and standard deviation of N_A, the number of inclusions in each field (edge corrections must be performed), also can be performed. Edge corrections are not required for the volume fraction but are needed for N_A, A and L.

Thus, with E 1245, particles can be characterized as to the V_V, N_A, A, L and λ values with means for each and standard deviations for all but λ (unless the constituent is present in every field). If a number of specimens per lot are tested, the mean and standard deviations for V_V, N_A, A, L and λ values for all of the specimens in the lot can be determined. These data can be entered into a spread sheet, much easier than JK-type data. Different lots can be compared statistically to detect any significant differences using standard statistical tests.

Examples of the use of E 1245 data to learn more about materials and processing have been published [12-15]. As an example of the method, and some important aspects of its use, Figures 1 to 4 show the volume fraction, number per mm^2, average area and mean free path of oxides (no sulfides were present) in specimens from an experimental VIM/ESR ingot of a Ni-based superalloy. Six specimens were taken from the as-cast ingot product, after electroslag refining, from six locations from top to bottom of the ingot. These

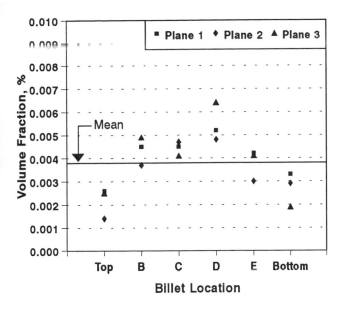

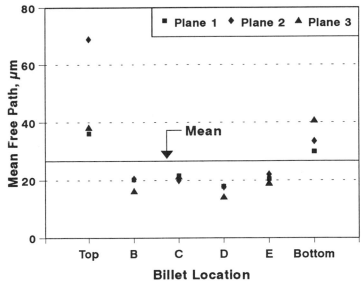

Fig. 1 - V_v of oxides in the experimental VIM/ESR ingot.

Fig. 2 - N_A of oxides in the experimental VIM/ESR ingot.

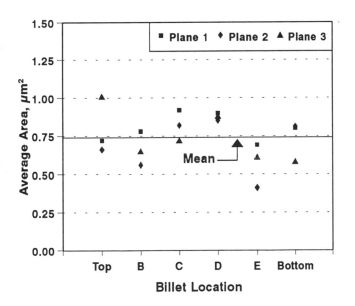

Fig. 3 - Average area of oxides in the experimental VIM/ESR ingot.

Fig. 4 - λ of oxides in the experimental VIM/ESR ingot.

specimens were hot-isostatically pressed to heal the shrinkage cavities which would exhibit the same gray level as the oxides. Each specimen was taken from the mid-thickness of the ingot. Image analysis E 1245 measurements were made with an 80X objective. Each specimen was polished and evaluated on three parallel test planes (repolished in between runs) with 200 fields per plane. Because the oxides were basically globular in shape, the length values are not shown.

Figure 1 shows the volume fractions for the three measurements on each of the six specimens recorded to four decimal places to better show the very minor differences at these very low oxide levels. Note that the volume fractions were much lower at the top and bottom of the ingot. We can compute the mean and standard deviation of the three measurements for each location and compare them to determine which differences are statistically significant at some chosen level (e.g., 95%). For the volume fraction data, there

27

was no statistical significance to the difference in volume fractions between the top and the bottom specimens, between the E location and the bottom and between locations B, C, D and E (the V_V difference between D and E was significant at the 90% confidence level). On the other hand, the V_V differences between the top location and locations B, C, D and E and between the bottom location and B, C and D were significant at the 95% confidence level.

Figure 2 shows that the N_A was much lower at the top and bottom compared to the interior. The N_A differences between the top and locations B, C, D and E and between the bottom and locations B, C, D and E were statistically significant. The N_A differences between the top and bottom locations and the N_A differences between the B, C, D and E locations were not significant.

Figure 3 shows that the average oxide cross sectional areas were quite similar for all locations. The only area differences that were statistically significant at the 95% confidence level were between locations B and D and D and E.

Figure 4 shows that, as might be expected, the mean free path (mean edge-to-edge spacing) were lower for the top and bottom locations where A_A and N_A were lower. Note that one λ reading (plane 2) for the top location had a much greater value causing the standard deviation to be high for this location. Locations B, C, D and E were highly repeatable. Statistically, the λ differences between the top location and the C location and bottom location were not significant at the 95% level. However, the λ difference between top location and the D location was significant while the λ differences between the top location and the B and E locations were significant at the 90% level. The λ differences between the bottom location and the D and E locations and between the D and C locations were significant while the λ difference between the D and E locations were significant at the 90% level. The λ differences between the top and the C location and between the C and E locations were not significant.

Consequently, the analysis showed that the oxides were fewer in quantity and more widely spaced apart, but of the same size, at the top and bottom compared to the interior of the ingot.

Table 2 summarizes the ingot averages, for each of the three planes (all six locations) and for all of the data. In the same way, we can test these differences. None of the differences, between results for each plane or between the grand mean and each plane, were significant even at the 80% confidence level. This demonstrates the repeatability of the test measurements, even at this very low inclusion level.

Grain Size Measurement

ASTM E 1382 [16] was written and approved in 1991 covering measurement of grain size using image analysis complementing ASTM E-4's long effort [17] to standardize manual test methods in E 112 [18].

In general, there has been less overall need for performing grain size measurements by image analysis compared to inclusion or second-phase particle assessment. For routine work, the chart estimation method is extremely fast and usually of sufficient precision, although there is bias in chart ratings of grain size [15]. For research work where it is necessary to determine the influence of grain size on some property, a precise, unbiased measurement is required. Manual measurements are satisfactory and simple to perform [19]. Characterization of duplex grain size distributions [20] is a good candidate for image analysis due to the need for extensive individual measurements of grain intercepts, diameters or areas.

ASTM E 1382 lists a number of approaches for measuring grain size by image analysis. Friel et al. [21] demonstrated that the influence of incompletely etched grain boundaries on precision and bias is less for intercept length measurements than for grain area measurements for the same number of fields and grains sampled. Dealing with twin boundaries in certain austenitic or face-centered cubic metals is always a problem, both for manual and automated measurements. While certain etchants can be used to suppress twin boundaries for some materials (as re-

Table 3. ASTM E 1245 Data for Experimental Ni-Base Alloy.

Data Set	V_V (%)	N_A (mm^{-2})	A (μm^2)	λ (μm)
Plane 1	0.0041	50.4	0.80	24355
Plane 2	0.0034	51.7	0.68	30393
Plane 3	0.0040	55.0	0.74	25019
All	0.0038	52.4	0.74	26572

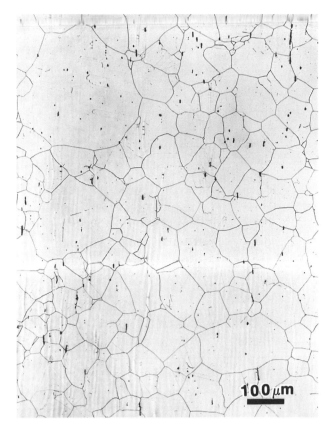

Fig. 5 - Grain structure of type 304 stainless steel specimen electrolytically etched with aqueous 60% nitric acid, 0.6 V dc, Pt cathode, 2 minutes.

viewed in [22]), this is not always possible. Friel and Prestridge [23] developed a procedure for identifying and removing twin boundaries from digitized images. Recently, Laroche and Forget [24] developed an algorithm for segmentation of color anodized grains in aluminum viewed with polarized light.

To illustrate some of the features of E 1382, data on the grain structure of a specimen of type 304 stainless steel [25], etched so that the annealing twins were not revealed, Figure 5, are presented. This structure was first assessed manually using the three-circle template described in ASTM E 112 [18]. Thirteen grid placements were made yielding 527 intercepts (207.3X magnification micrograph), an average intercept length of 59.5 ± 4.6 μm, and an ASTM grain size of 4.85 (G). The planimetric method of E 112 was also employed, assessing 8 fields, yielding on N_A of 230.1 grains/mm^2 (4346 μm^2 average area), and an ASTM grain size of 4.9.

E 1382 describes a number of approaches for determining the grain size. One approach is to determine the area of a number of grains, compute an average grain area, and relate this to G. Generally, the grain boundaries are detected, completed (missing portions), thinned, and the image is inverted revealing the grain interiors which are measured by the image analyzer. If a digitizing tablet is used, the grains are traced and the area is determined. The type 304 stainless steel micrograph was evaluated using a digitizing tablet with 180 grains traced. The average grain area was 4091.8 μm^2 which correlates to an ASTM grain size of 4.98 (G).

Since the cross-sectional areas of 180 grains have been measured, a plot of the distribution of grain areas can be constructed. A larger population would be preferred but 180 is adequate to reveal the grain area distribution. There are a number of ways that a grain area distribution can be plotted and some are better than others.

The simplest approach is to segregate the grain areas into a number of uniformly sized bins using a linear scale. The number of bins must be chosen. In general, the number of size classes must not be too small, e.g., less than 7, nor too large, e.g., not more than 15. As the number of classes increases, more grains must be measured to obtain a decent approximation of the distribution.

Figure 6 shows the simplest approach, the number percent of grains per size class using 13 classes with a linear scale. The vertical line is the mean grain area, 4091.8 μm^2. This plot is skewed right, that is, it is affected by a higher than normal number of large grain areas. This is also shown by the calculation of β_1 by the four-moment method [3] where a positive value indicates a right skewed distribution (for a normal distribution, β_1 = 0). Furthermore, β_2 (kurtosis) is very high indicating a leptokurtic distribution (for a perfect normal distribution, β_2 = 3; a value >5 indicates that the distribution is not Gaussian). Figure 6 suggests that the distribution may be better described by a lognormal distribution, although a small degree of duplexity is present (a bimodal distribution).

Figure 7 shows the result of grouping the logarithmic values of the grain areas into uniformly sized class intervals. Class one, for example, contains the number percent of all the grains with logarithms of their areas between 1 and 1.5, centered on 1.25 (17.8 μm^2). Class two contains the number percent of all the grains with logarithms of their areas between 1.5 and 2, centered on 1.75 (56 μm^2), and so forth. Note that Figure 7 is a bit closer in shape to a normal curve, indicating that the distribution is much better described by the log-normal distribution curve. Also, β_1, the skew, is very slightly negative, that is, left skewed, and β_2, the kurtosis (2.76), is very close to the ideal value of 3 (four-moment method applied to the log values).

In Figures 6 and 7, the frequency was calculated based on the number per class. Note that the difference between the mean value and the mode is less for the logarithmic distribution. Because grain areas are being measured, it may be better to calculate the area fraction of grains in each size class, as shown

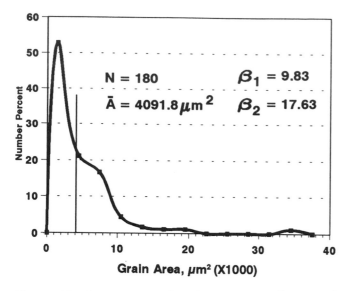

Fig. 6 - Number percent of grain areas on a linear scale.

Fig. 7 - Number percent of grain areas on a log scale.

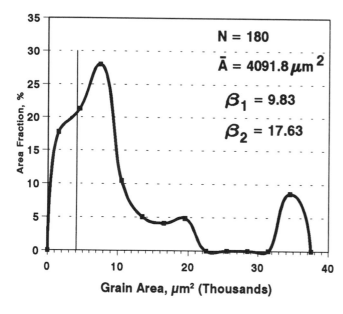

Fig. 8 - Area percent of grain areas on a linear scale.

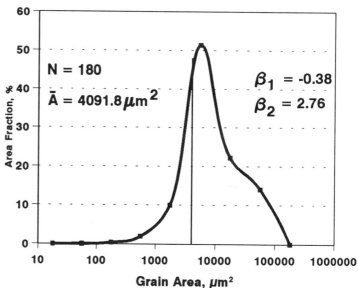

Fig. 9 - Area percent of grain areas on a log scale.

in Figures 8 and 9 for the linear and logarithmic distributions shown in Figures 6 and 7. This does not change the calculation of β_1 and β_2, of course. However, an area percent per class would, logically, appear to be a better way to graph the distribution of grain areas. Figure 8, the area fraction of grains using a linear scale, emphasizes the presence of the larger grains, which, while relatively few in number (not really seen in Figure 6), cover a fair percentage of the surface area and are obvious to the micrograph viewer.

Figure 9 shows the area percent of grain areas per class using a logarithmic scale. Note that, except for the

bulge on the right side of the distribution curve, for grains larger than 11,405 μm^2, the distribution curve is very good. Figure 9 shows the effect of the duplexity (two overlapping grain size distributions - too many very coarse grains) much better than Figure 7 (which makes it appear that there are too many fine grains).

The logarithmic classification approach appears to be more useful than the linear classification. There is another way to bin the grain areas in logarithm fashion that offers value because it is based upon the ASTM grain size scale. G is based on an exponential power series:

Table 4. Classification by ASTM G.

ASTM Grain Size No. (G)	Range of Grain Areas (μm^2)
00	182,412 - 364,796
0	91,239 - 182,412
1	45,620 - 91,239
2	22,810 - 45,620
3	11,405 - 22,810
4	5703 - 11,405
5	2851 - 5703
6	1426 - 2851
7	713 - 1426
8	356 - 713
9	178 - 356
10	89.1 - 178
11	44.6 - 89.1
12	22.3 - 44.6
13	11.1 - 22.3
14	5.6 - 11.1
15	2.8 - 5.6

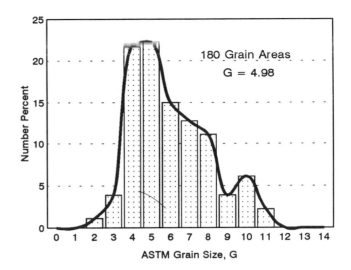

Fig. 10 - Number percent of grains on the ASTM scale.

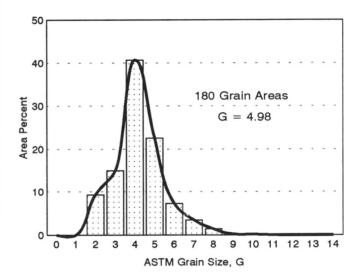

Fig. 11 - Area percent of grains on the ASTM scale.

$$n = 2^{G-1} \qquad (2)$$

where n is the number of grains per in^2 at 100X (15.5 times the number of grains per mm^2, N_A. at 1X). The reciprocal of N_A is the average grain area, A. Thus, for each grain size number, there exists a corresponding mean grain area. Thus, we can determine the number percent of grains or the area percent of grains, with an area typical of a specific G value. For example, all grains with an area between that for G = 2.5 and G = 3.5 will be of grain size 3, and so forth. Table 4 lists the range of grain areas for G values between 00 and 15.

The 180 grain areas were classified in this manner and the number percent and area percent per class, as a function of the ASTM grain size number, are shown in Figures 10 and 11. While this appears to be a linear classification, it is a logarithmic classification since the width of each area class is equal to the log of 2. The β_1 and β_2 calculations do not change because we are still actually dealing with the logarithms of the grain areas. It is quite clear that the area percent plot, Figure 11, is an excellent depiction of the distribution, very much like Figure 9 except that the X axis is reversed. It is clear in Figure 11 that the distribution is close to a log-normal distribution but contains more grains of ASTM 2 than it should, because of the slight degree of duplexity.

In the exact same manner, the distribution of intercept lengths can be handled. They can be grouped and plotted in the same manner as shown in Figs. 6 to 9. Or, intercept length limits can be calculated, as was shown in Table 4 for grain areas, to classify intercept lengths (chords cutting the grain areas at random) to ASTM G values. The scale increases approximately as the log of the square root of 2. Plots similar to those shown in Figs. 10 and 11 can then be constructed. The area percent (and the lineal percent for intercept lengths) is preferred to the number percent because the frequencies described relate directly to the physical entity being measured, that is, area (or length). The ASTM logarithmic plot, compared to a standard logarithmic scale (that is, Fig. 11 vs Fig. 9), has the advantage that ASTM grain size numbers are more understandable to the metallographer than square micrometre range values.

Summary

The development of three standard test methods employing image analysis to characterize micorstructures - ASTM E 1122, E 1245 and E 1382 - has been reviewed. These methods are a vast improvement over earlier chart based methods because they are quantitative, rather than qualitative. The use of image analysis improves repeatability and reproducibility. Test methods must be clearly defined and logically developed if they are to become useful standards.

References

1. Vander Voort, G.F., ASTM Standardization News, 19, 58-77 (May 1991).
2. Vander Voort, G.F., ASTM STP 1165, Metallography: Past, Present, and Future (75th Anniversary Volume), ASTM, Philadelphia, 3-79 (1993).
3. Vander Voort, G.F., Metallography as a Quality Control Tool, Plenum Press, N.Y., 1-88 (1980).
4. Rinman, B., H. Kjerrman and B. Kjerrman, Jernkontoret Ann., 120, 199-226 (1936).
5. ASTM E 45, Standard Test Methods for Determining the Inclusion Content of Steel.
6. ISO 4967, Steel - Determination of Content of Non-Metallic Inclusions - Micrographic Method Using Standard Diagrams.
7. Vander Voort, G. F. and R. K. Wilson, ASTM Standardization News, 19, 28-37 (May 1991).
8. Vander Voort, G. F. and J. F. Golden, Micorstructural Science, 10, Elsevier Science Publishing Co., 277-290 (1982).
9. ASTM E 1122, Standard Practice for Obtaining JK Inclusion Ratings Using Automatic Image Analysis.
10. Forget, C., ASTM STP 1094, MiCon 90: Advances in Video Technology for Microstructural Control, ASTM, Philadelphia, 135-150 (1991).
11. ASTM E 1245, Standard Practice for Determining the Inclusion or Second-Phase Constituent Content of Metals by Automatic Image Analysis.
12. Vander Voort, G. F., ASTM STP 987, Effect of Steel Manufacturing Processes on the Quality of Bearing Steels, ASTM, Philadelphia, 226-249 (1988).
13. Vander Voort, G. F., Inclusions and their Influence on Material Behavior, ASM Intl., Metals Park, OH, 49-64 (1988).
14. Vander Voort, G. F., Materials Characterization, 27, 241-260 (1991).
15. Vander Voort, G. F., Quantitative Microscopy and Image Analysis, ASM Intl., Materials Park, OH, 21-34 (1994).
16. ASTM E1382, Standard Test Methods for Determining Average Grain Size Using Semiautomatic and Automatic Image Analysis.
17. Vander Voort, G. F., ASTM Standardization News, 19, 42-47 (May 1991).
18. ASTM E 112, Standard Test Methods for Determining Average Grain Size.
19. Vander Voort, G. F., ASTM STP 839, Practical Applications of Quantitative Metallography, ASTM, Philadelphia, 85-131 (1984).
20. Vander Voort, G. F. and J. J. Friel, Materials Characterization, 29, 293-312 (1992).
21. Friel, J. J., E. B. Prestridge and F. Glazer, ASTM STP 1094, MiCon 90: Advances in Video Technology for Microstructural Control," ASTM, Philadelphia, 170-184 (1991).
22. G. F. Vander Voort, Metallography: Principles and Practice, McGraw-Hill Book Co., N.Y., 235, 238-239 (1984).
23. Friel, J. J. and E. B. Prestridge, ASTM STP 1165, Metallography: Past, Present, and Future (75th Anniversary Volume), ASTM, Philadelphia, 243-253 (1993).
24. Laroche, S. and C. Forget, Proceedings of Metallography Congress '95, Colmar, France, to be published.
25. Vander Voort, G. F., ASTM STP 1165, Metallography: Past, Present, and Future (75th Anniversary Volume), ASTM, Philadelphia, 266-294 (1993).

An Experimental Method for Quantitative Characterization of Spatial Distribution of Fibers in Composites

S. Yang, A. Tewari and A. M. Gokhale
Georgia Institute of Technology, Atlanta, GA

S. Yang, A. Tewari, and A.M. Gokhale
Department of Materials Science and Engineering
Georgia Institute of Technology
Atlanta, Georgia-30332-0245

Abstract

A practically feasible experimental procedure is presented for statistically robust and unbiased quantitative characterization of spatial arrangement of fibers in unidirectionally aligned fiber composites. The application of the procedure is demonstrated via estimation of nearest neighbor distribution, K-function, and radial distribution function of non-uniformly distributed Nicalon (SiC) fibers in a glass ceramic matrix composite.

FIBER REINFORCED COMPOSITES are an important class of advanced materials where the properties and performance critically depend on microstructure. Spatial arrangement of fibers is an important microstructural attribute in the fiber composites. The damage evolution and mechanical behavior of composites depend significantly on the spatial distribution of fibers. For example, it has been experimentally shown that propensity of micro-cracking during thermal cycling of metal matrix composites depends on the distances between fibers[1,2], and the transverse failure stress depends on the spatial distribution of fiber centers[3,4]. Therefore, it is of interest to quantitatively characterize the spatial distribution of fibers in composites.

In the past, very little attention has been given to the quantitative characterization of spatial arrangement of fibers in composites due to lack of suitable characterization methods, and also lack of theoretical models that can utilize such detailed quantitative data. However, recently developed computer simulation and analytical models consider deviations from uniformity and periodicity of spatial distribution of microstructural features such as fibers. These studies have demonstrated the effect of spatial distribution of fibers/particles on the damage evolution and mechanical response of composites. For example, Brockenbrough et al.[5] have shown that stress-strain behavior of metal matrix composites depends significantly on the spatial arrangement of fibers/particles. Sorenson and Talreja[6] have shown that local residual stresses during cool-down during processing of composites is very sensitive to deviations of spatial distribution of fibers from uniformity. They have also shown that[7] the effectiveness of coatings on fibers depends on the non-uniformities in the spatial distribution of fibers. Thus, the theoretical studies are now reaching the level of sophistication where detailed quantitative data on the spatial arrangement of fibers can be fruitfully utilized.

Statistically reliable quantitative characterization of spatial arrangement of fibers requires measurement of location (as quantified by centroid co-ordinates) and sizes of large number of fibers, and therefore automatic digital image analysis is <u>required</u>. There are three major difficulties associated with statistically reliable and unbiased quantitative estimation of the descriptors of spatial arrangement of fibers using automatic image analysis: (1)How to account for the fibers that are on the edges of the field of view, and therefore their centers may not be in the field of view under observation (see Figure-1a), (2) How to measure distances between fibers that are not in the same field of view (see Figure-1b), and (3) Image analyzer regards a string of touching fibers as one feature, and therefore how to automatically detect individual fibers (and fiber centers) when fibers are very closely spaced or touching one another (see Figure-1c). Recently, the first two difficulties have been resolved by developing an image analysis procedure[8] for creating a "montage" of large number of adjoining perfectly matching microstructural fields in the computer memory, which eliminates the "edge effect" for all the

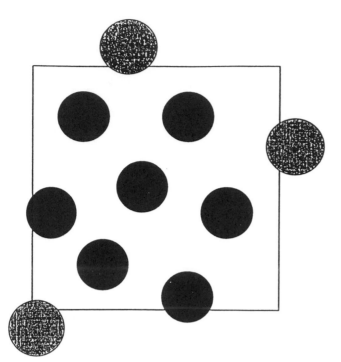

Figure-1a A typical microstructural field, where the centroids of some features(grey circles) are outside the frame boundary

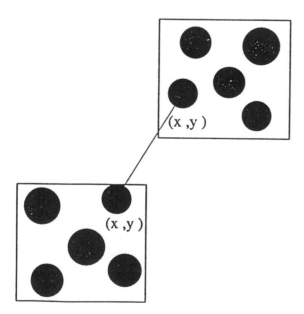

Figure-1b Estimation of distance between fibers that are in different fields of view

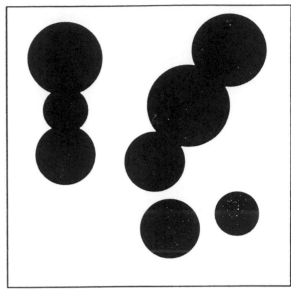

Figure-1c Closely spaced particles may be treated as a single feature by image analyzer

practical purposes, because edges are eliminated! In the present contribution, this image analysis procedure is used for quantitative estimation of important descriptors of spatial arrangements of fibers in a glass ceramic matrix composite. Further, a simple image analysis algorithm[9], called "grain process" is used to automatically identify and accurately measure centroid coordinates and sizes of individual fibers in a population where many fibers are very closely spaced or touch one another. The combination of the two techniques, namely, "montage" creation, and automatic individual fiber identification using grain process, establishes a practical experimental procedure for unbiased and statistically reliable quantitative characterization of spatial arrangement of fibers in unidirectional composites. The image analysis yields the (X, Y) centroid coordinates and sizes of individual fibers in a large number of contiguous microstructural fields. All the centroid coordinates are referred to the same **origin**, and therefore distance between any two fibers (whose centroids may or may not be in the same field of view) can be easily calculated. In the present paper, these basic image analysis data are utilized to compute the descriptors of spatial distribution such as nearest neighbor distribution and radial distribution function for the population of Nicalon (SiC) fibers in a glass ceramic matrix composite. It is demonstrated that the spatial distribution of fibers in highly non uniform. These spatial distribution data can form a realistic input to model a representative volume element (RVE) for computer simulation studies on damage evolution and mechanical response of ceramic matrix composites.

A brief background on important descriptors of spatial distribution of fibers is given in the next section, and this is followed by description of the image analysis procedure and computer calculations for estimation of the descriptors of the spatial distribution. The experimental procedure and data are presented in a subsequent section, which is followed by the interpretation of the spatial distribution data and discussion.

Background

The distribution of relative locations of fibers is manifested in the spatial patterns, correlations, clustering, short and long range interactions, segregation, etc. The spatial characteristics include descriptors such as radial distribution function, nearest neighbor distribution function, 2nd, 3rd, ---, nth neighbor distribution functions, etc. A brief review of these important descriptors of spatial order is given below.

K-function[8,10-12], radial distribution function[8,10-12], nearest neighbor distribution function[13-15], 2nd, 3rd, ---, nth neighbor distributions[16,17], etc. are useful to characterize general spatial clustering, repulsion, or randomness of the population of fibers. Consider microstructure of a composite containing unidirectional aligned fibers, as observed in a metallographic plane perpendicular to fibers. Let A_0 be the area of the metallographic plane, and let N be the number of fiber centers in this plane. The **average** number of fibers per unit area N_A is obviously equal to (N/A_0). Draw a test circle of radius r around the center of a fiber at (X_i, Y_i) and count the number of other fiber centers in this test circle. Repeat this measurement for all the N fibers(i.e., i = 1, 2, ---, N) and calculate the **average** number of other fiber centers in a test circle of radius r drawn around a typical fiber from these data; this average $K_A(r)$ is called K-function of the fibers[8,10-12,16,17]. Calculate the K-function for different distances r. For completely random arrangement of point particles (i.e., zero size), $K_A(r)$ is equal to $(\pi r^2).N_A$.

The radial distribution function $g_A(r)$ is related to the derivative of the K-function, and it is defined as follows[8,10-12,16,17].

$$g_A(r) = \frac{1}{2\pi r \cdot N_A} \cdot \frac{dK_A(r)}{dr} \quad ----(1)$$

$g_A(r)$ can be experimentally measured in a metallographic plane perpendicular to the fibers. The radial distribution function may be interpreted as the ratio of the average number of fiber centers in a circular shell of radii r and (r+dr) around a typical fiber and the corresponding number for completely randomly distributed points in a plane having the same number density N_A. For randomly distributed points $g_A(r)$ is equal to one; a value significantly higher than one represents clustering, and value lower than one represents repulsion. To obtain statistically reliable estimates of the K-function or the radial distribution function, it is necessary to perform the measurements on at least five hundred fibers, and therefore automatic image analysis is essential. **It is necessary to point out that for values of r larger than half the size of microstructural field, the radial distribution function or K-function cannot be estimated from individual disconnected fields of view, and therefore it is essential to create a montage of large number of perfectly matching contiguous microstructural fields by appropriate image analysis procedure.**

The nearest neighbor distribution function[13-15] is given by the probability density function $\psi(r)$ such that $\psi(r)dr$ is equal to the probability that there is no other fiber center in a circle of radius r around a typical fiber, and there is at least one fiber center in the circular shell of radii r and (r+dr). The nth nearest neighbor distribution function $\psi_n(r)$ is the probability density function such that $\psi_n(r)dr$ is equal to the probability that there are exactly (n-1) other fiber centers in a circle of radius r around a typical fiber, and at least on fiber center in the circular shell of radii r and (r+dr). To obtain reliable estimates of the nearest neighbor and higher order neighbor distribution functions, the measurements must be averaged over at least five hundred fibers. **The nearest neighbor (or a higher order neighbor) of a given fiber may NOT be in the same microstructural field, and therefore these distribution functions cannot be estimated from measurements performed on disconnected and separate individual micrographs or fields of view!** Therefore, it is necessary to create a montage of large number of perfectly matching adjoining fields of view for reliable estimation of nearest neighbor distribution and higher order neighbor distribution functions through appropriate image analysis procedure.

Image Analysis

As explained in the previous section, for reliable and unbiased measurements of the descriptors of spatial arrangement of fibers, it is necessary to create a montage of large number of perfectly matching microstructural fields in the memory of the image analysis computer. Creation of the image montage eliminates the edge effect for all the practical purposes, it facilitates measurement of distances between fibers that are in different fields of view, and it enables the estimation of the K-function and radial distribution functions for distances larger than the microstructural field size.

In the present approach, the montage of contiguous microstructural fields is created by using the image analyzer in an interactive mode. The first field of view (FOV) is arbitrarily taken at the place of concern and stored in the computer memory. The right border (of about 60 pixel width) of this image is stored on the left edge of a blank image. The resultant semi blank image is then displayed along with the live image. This results in a superimposed image on the left border of the screen (of the previous right border and live image) with rest of the screen having the live image. The microscope stage is then manually moved so that the right border of the live image moves to the left border and they are approximately matched. The perfect matching is achieved through pixel by pixel shifting of the live image. This results in a match of the first and the second image with an accuracy of one pixel. This second image is then stored in the computer memory. All successive images are grabbed by using the same procedure to create a continuous montage of microstructural fields .

In the fiber composites, the volume fraction of fibers and their number density are often high. As a result, the fibers are very closely spaced or touching one another. This together with the limited resolution of microscope, and limited grey levels lead to binary images in which strings of fibers are connected to one another. In such cases, it is first necessary to "separate" the fibers, before the centroid coordinates of individual fibers can be measured. This can be achieved by using a number of different image processing algorithms. The most popular algorithm involves pixel by pixel "erosion" of the binary image. The "erosion" shrinks the fibers, and in the process the connecting bridges between the adjacent fibers shrink as well. As the connecting bridges are very small as compared to the fibers, after few erosions all the fibers get separated in the binary image. The process separates the touching fibers, but it results in the images of the fibers having sizes smaller than the original ones. This is corrected by an image operation called "dilation". The dilation expands the fibers step by step, keeping their separate identifications. This procedure is routinely used for image segmentation (i.e., to identify each fiber as a separate feature). This process does achieve fiber separation, but it leads to two errors:(1) very small fibers shrink to zero size during erosion, and therefore they are "lost", and can not be regenerated during dilation, and (2) the process of regeneration (dilation) cannot achieve the same exact shape and size of the fibers as in the initial binary image. Due to these limitations, this popular algorithm is not utilized in the present work.

In the present approach, the fiber separation is achieved by an alternate algorithm, based on "grain process". In this procedure, the initial binary image is stored in a buffer. The grain process is then applied to the image (without changing the buffer). The dilation is carried out beyond the point where the same size is achieved. This results in fibers of size bigger than their original size, but the identity of separate fibers is still kept. This separate identity is used to draw one pixel thick lines to separate different features. The resulting image contains fibers of a bit bigger size than the original, separated with one pixel thick lines. This image is than combined with the initial image stored in the buffer using Boolean AND to get the final image. This final image thus has fibers of the same size and shape as original along with one pixel thick line of separation between different fibers, and therefore the touching fibers are detected as separate features.

Once a continuous montage of FOV is stored and the fibers are separated, the final step is to extract the microstructural data out of it. The main problem is posed by the image analysis systems which can measure only one screen at a time. Therefore, to again get around the edge effect problem, the first field of view (FOV) is displayed on the screen (*analysis screen*). The measurements are done on a frame of size one forth of the analysis screen (known as *measurement frame*). The measurement frame is taken at the center of the analysis screen so that the fibers at the edge of the analysis screen are not needed to be accounted. The next analysis screen is displayed by taking half of the first FOV and half of the second FOV. The measurement is again done on a measurement frame (centered on the screen) of size one fourth of that of the analysis screen. Similarly all other FOV are analyzed using this procedure. By doing this all the fibers present in the montage are measured and the same fiber is never measured twice. This algorithm is described in detail elsewhere[8].

The output of the image analysis program gives the coordinates of the centroids of every fiber referred to a common origin, and the parameters such fiber size, shape, aspect ratio, etc. of each fiber. These data contain all the information present on the montage (composite field). Thus once this data is extracted from the montage all the measurements and calculations can be done through these data without going back to the montage. This data set is referred as microstructural data in the subsequent sections.

Computer Calculations

The K-function of the fiber centers can be estimated from the microstructural data by using the following algorithm.

(1) Draw a circle of radius Δr around the centroid of an arbitrarily selected fiber.

(2) Count the number of other fiber whose centroids lie inside the circle of radius Δr. This amounts to counting the number of fibers whose centers are at a distance $d \leq \Delta r$ from the center of the reference fiber.

(3) Repeat step no. (2) by making each fiber, the reference fiber turn by turn.

(4) Calculate the average of all these counts. The average is the value of K function for that specific value of distance equal to Δr.

(5) Repeat steps 1 to 5 for distances $2\Delta r$, $3\Delta r$,-,-,$n\Delta r$ and so on to obtain the value of K-function for different distances.

(6) The above steps can continue till the circle drawn around a fiber becomes so large that it intersects the boundary of the montage. At this stage there are two choices, first to terminate the calculation, the second is to go on doing the calculation by linear extrapolation for the incomplete part of the circle.

The K-function can be estimated from the microstructural data set in this manner. Once the K-function is known, the radial distribution function, $g_A(r)$ can be calculated by using equation (1) via numerical differentiation of the K-function.

Measurement of the nearest neighbor distribution function is much easier than the measurement of K-function or radial distribution function. Distance of the nearest neighbor of each fiber is calculated from the microstructural data set. Distribution of these distances are presented as the nearest neighbor distribution function. Fibers which are very close to the edge (i.e. fibers which are

closer to the edge than their nearest neighbor present in the montage) are not considered in this function. Similar procedure can be used to calculate the higher order neighbor distributions.

Experimental work on ceramic matrix composite
1. Material.

The measurements were performed on glass ceramic matrix composite (Nicalon/MAS-5), supplied by Corning Incorporated. The composite consists of MAS cordierite($2MgO \cdot 2Al_2O_3 \cdot 5SiO_2$) matrix and Nicalon (SiC) fibers. Uni-ply of continuous fibers are stacked together to get the required geometry of the composite. The fibers are unidirectionally aligned in the matrix. The fibers have an average diameter of 14 µm and their volume fraction is 0.35 (i.e., 35%). The average number of fibers per unit area, N_A is equal to 2260/mm². Figure-2 depicts the size distribution of the fibers in this composite. The fiber diameters vary from 3 to 29µm. However, most of the fibers are in the range of 11-17 µm. The fiber size distribution is approximately normal.

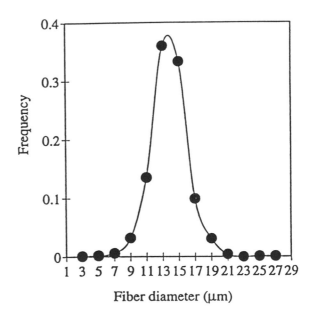

Figure-2 Size distribution of Nicalon fibers in the composite

2. Metallography and Microstructure of Composite.

To characterize the spatial distribution of fibers, the samples were cut to expose a randomly located metallographic plane perpendicular to the fibers. A lot of care was taken to minimize damage to the fibers during cutting of the specimens. After cutting, the samples were mounted, ground and polished by using the standard metallographic procedures. Figure-3 shows the microstructure revealed in this manner at three different magnifications. Observe that all the fibers are not of the same size, and they are non-uniformly distributed in the matrix. In some regions, the fibers are heavily clustered, which gives rise to fiber rich regions. There are some regions where there are only a few fibers or no fibers at all. These regions are the so called repulsion regions or "fiber depleted" regions. The mechanical and thermal cycling response of such nonuniform microstructure is expected to be sensitive to the spatial arrangement of the fibers. Therefore, it is useful to quantify the spatial arrangement of fibers in such materials.

3. Digital Image Analysis.

To estimate the descriptors of spatial arrangement of fibers, the image analysis work was performed on VIDAS image analysis system supplied by Carl Zeiss, Inc. A "montage" of fifteen perfectly matched contiguous microstructural fields was created in the memory of the image analysis computer by using the algorithm described in **Image Analysis**. Two montages were generated for the quantitative measurements; each montage contained more than five hundred fibers. The coordinates of the fiber centers in each montage, the size of each fiber, as well as the shape factor of the fiber were measured automatically. From the microstructural data set obtained from the image analysis, the microstructure of the composite can be reconstructed as shown in Figure-4

The K-function, nearest neighbor distribution function and the radial distribution function were calculated by applying the above devised computer programs. The results of these calculations are presented in the following section.

Results and Discussion

K-function for the spatial distribution of fibers is given in Figure-5, and the corresponding radial distribution function is given in Figure-6. The radial distribution function is presented in the normalized form; the distance r is normalized by the average fiber radius $\bar{r}$. In this manner, the radial distribution is normalized by the microstructural scale of the Nicalon fibers. In Figure-6, observe that the radial distribution function $g_A(r)$ is zero for the distances less than two times the average fiber radius, it then quickly peaks to a high value, and then oscillates and settles down to a value close to one.

Recall that the radial distribution function

$g_A(r)$ is equal to the ratio of the number of other fiber centers in a circular shell of radii r and (r+dr) around a typical fiber, and the corresponding number for randomly distributed fibers of zero size. Then the radial distribution function is equal to one for all the distances if the fibers having zero size are randomly distributed. The experimentally measured radial distribution function may differ from one (i.e. that for randomly distributed zero size fibers) because of two reasons: (i) deviation of the spatial distribution of fiber centers from randomness, and/or (ii) finite size and volume fraction of fibers. It is not possible to deconvolute the effects of these two variables on the radial distribution function in a straight forward manner. One way to resolve this problem is to compare the experimentally measured radial distribution function with the radial distribution function for a **computer simulated** microstructure where fibers have same size distribution, number density, and volume fraction, **but there is random spatial distribution of fibers**. Figure-7 shows the normalized radial distribution function for the computer simulated microstructure, where the fibers are randomly positioned, but they have the same size distribution, number density, and volume fraction as in the experimentally characterized microstructure. Figure-8 depicts this simulated microstructure. Comparison of Figures 6 and 7 reveals the following.

(1) The radial distribution function for both the microstructures is zero for $r < 2\bar{r}$. This basically conveys that due to the effect of the finite fiber size, on the average, the centers of two fibers can not be at a distance less than the average fiber diameter from one another, as expected. Therefore, for $r < 2\bar{r}$, g(r) is zero due to the finite size effect.

(2) Both of the radial distributions show first peak at a distance somewhat larger than the average fiber diameter. In general, the position of the first peak depends on the average fiber size and the fiber size distribution[18]. However, the height of the first peak is significantly larger for the experimentally measured radial distribution function (1.93) as compared to that for the computer simulated structure where the fibers are randomly positioned. This shows that in the present glass ceramic matrix composite, the fibers are **not** randomly distributed **and** there is short range clustering of fibers. The ratio of the height of first peak of the experimental radial distribution function and that for the simulated structure (1.93 / 1.4 ~ 1.4) is one quantitative measure of the short range deviation of the spatial fiber distribution from the randomness.

(3) At large distances, the experimental radial

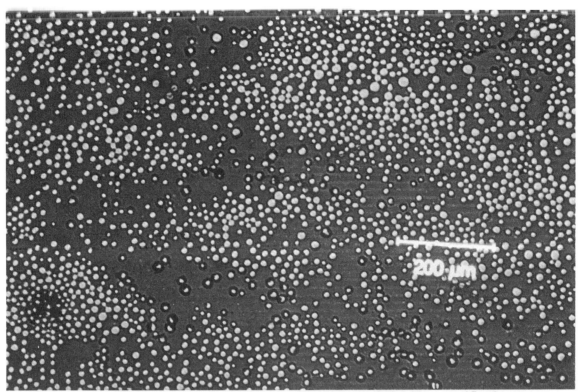

Figure-3a Microstructure of the composite revealed at low magnification.
Note that there are fiber rich and fiber poor regions

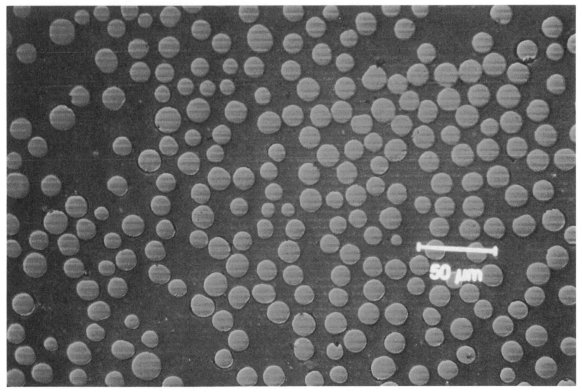

Figure-3b Same microstructure as in Figure-2a, but at somewhat
higher magnification

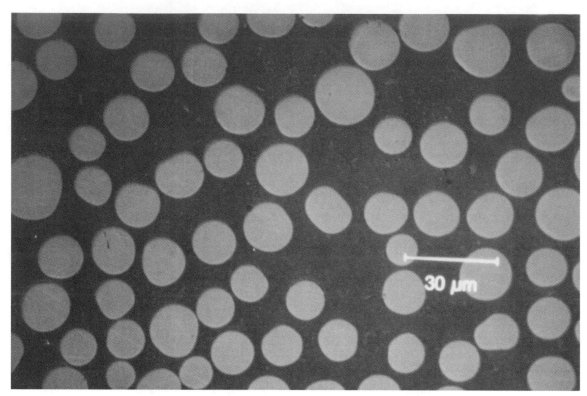

Figure-3c Microstructure of the composite revealed at high magnification

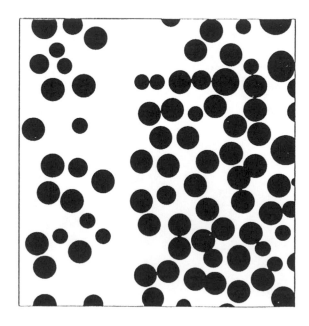

Figure-4 Microstructure of the composite regenerated from the microstructural data set

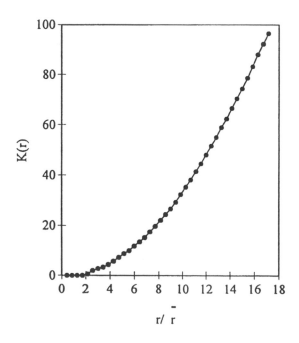

Figure-5 K-function of the Nicalon fibers in the composite

40

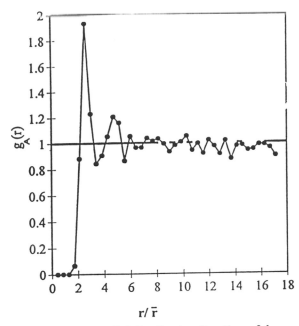

Figure-6 The radial distribution function of the Nicalon fibers

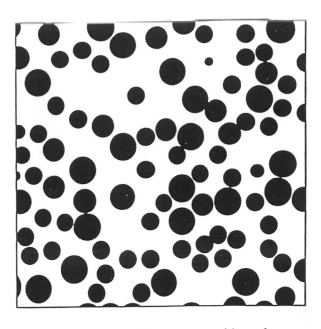

Figure-8 Simulated microstructure with random fiber distribution whose radial distribution function is given in Figure-7

distribution approaches a mean value that is **less than 1.0,** but at the comparable large distances the radial distribution function for the simulated structure has a mean value equal to 1.0. This shows that in the microstructure of the composite there is "scarcity" of fibers at large distances (long range). *It is concluded that the microstructure of the composite consists of fiber rich and fiber poor regions.* This is quite consistent with the nature of microstructure shown in Figure-3. The ratio of the height of the first peak of the experimental radial distribution to that of the simulated structure is a quantitative measure of the short range clustering of the fibers, and therefore it describes the "intensity" of the fiber rich regions. The scarcity of the fibers in the fiber poor regions is quantified by the deviation of the mean value of the radial distribution function at **large** distances (say, at distances in the range of $8\bar{r}$ to $16\bar{r}$) from 1.0. For the present composite, this deviation is equal to 0.03.

Figure-9 shows the experimentally measured nearest neighbor distribution function of the Nicalon fibers in the present glass ceramic matrix composite. The nearest neighbor distribution function depends on the spatial arrangement of fibers as well as the fiber size, size distribution, and volume fraction. Therefore, useful information about the spatial arrangement of fibers can be obtained from the nearest neighbor distribution function only through its comparison with the corresponding distribution for a simulated microstructure having randomly

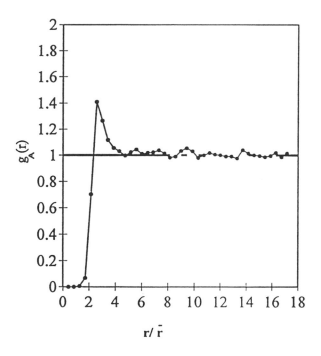

Figure-7 The radial distribution function for computer simulated microstructure having random spatial distribution of fibers

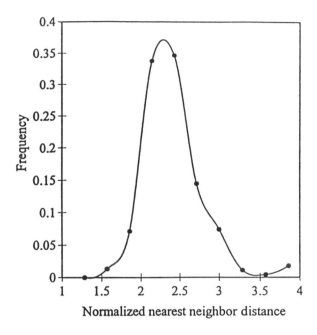

Figure-9 The nearest neighbor distribution function of the fibers in the composite

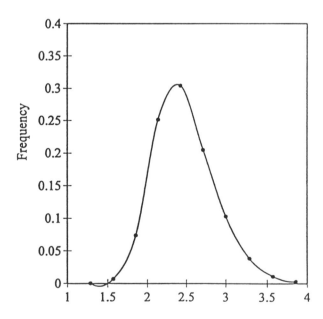

Normalized nearest neighbor distance

Figure-10 The nearest neighbor distribution function for the computer simulated microstructure having random fiber distribution

located fibers, but the same fiber size distribution and volume fraction. The nearest neighbor distribution function for the computer simulated microstructure having randomly located fibers of same size distribution and volume fraction is shown in Figure-10. The average nearest neighbor distance between the Nicalon fibers can be calculated from the nearest neighbor distribution shown in Figure-9, and it is equal to $17\mu m$. Note that this value is obtained from the measurements on more than one thousand fibers. The average nearest neighbor distance for the computer simulated microstructure (Figure-10) is also approximately equal to $17\mu m$. Therefore, the nearest neighbor distance of nonuniformly distributed fibers in the composite is close to that of the randomly distributed fibers. Furthermore, the nearest neighbor distribution function of the composite is quite similar to that of the simulated random microstructure. This is not surprising because the nearest neighbor distribution function only reflects the very short range spatial distribution of features and the simulated microstructure has approximately the same average size, same size distribution, and same number density as the composite.

Due to the nature of its definition, the nearest neighbor distribution contains information about the *short range* arrangement of fibers. However, it does not provide information concerning *long range* scarcity of fibers or long range deviations from randomness., On the other hand, the radial distribution function provides information about both the short range and long range spatial distribution of the fiber locations. *Radial distribution function and the nearest neighbor distribution function are independent descriptors that contain information about different facets of the spatial arrangement of fibers.* The nearest neighbor distance appears as a parameter in some models for structure property correlations as well as in computer simulation studies. The present contribution provides an experimental procedure for statistically reliable and unbiased estimation of the nearest neighbor distance.

Conclusions

An experimental procedure is developed for quantitative characterization of spatial arrangement of fibers in unidirectional fiber composites. The procedure involves digital image analysis. Any descriptor of the spatial arrangement of fibers can be computed from the resulting image analysis data. This technique is successfully applied to quantify the spatial arrangement of Nicalon fibers in a glass ceramic matrix composite. It is concluded that the microstructure of the composite consists of fiber rich and fiber poor regions. The extent of the fiber clustering in the fiber rich regions, and the "scarcity" in the fiber poor regions can be quantified by using the radial distribution function.

Acknowledgement

This research is supported by National Science Foundation sponsored project (DMR-9301986) "Quantitative Analysis of Fracture Surfaces Using Stereological Methods". Dr.B. MacDonald is the project monitor. The financial support is gratefully acknowledged.

References

(1) A.M. Gokhale and W.T. Whited, "Measurements of Growth Rates of Thermally Induced Microcracks in a Metal Matrix Composite", in **Developments in Ceramic and Metal Matrix Composites**, K. Upadhya, ed., TMS, Warrendale, PA., 1992, PP. 273-286.

(2) W.T.Whited, A.M. Gokhale, and N.U. Deshpande, "Characterization of Thermally Induced Microcracks in Metal Matrix Composite", **Microstructural Science**, 1994, Vol. 21, PP.107-120.

(3) R. Pyrz, "Correlation Between Microstructure Variability and Local Stress Field in Two-Phase Materials", **Materials Science and Engineering**, 1994, Vol. A177, PP. 253-259.

(4) R. Pyrz, "Interpretation of Disorder and Fractal Dimension in Transversely Loaded Unidirectional Composites", **Polymer and Polymer Composites**, 1993, Vol.1, PP. 283-295.

(5) J.R. Brokenbrough, S. Suresh, and H.A. Wienecke, "Deformation of Metal Matrix Composites with Continuous Fibers: Geometrical Effects of Fiber Distribution and Shape", **Acta Metall. Mater.** 1991, Vol.39, PP. 735-752.

(6) B.F. Sorensen and R. Talreja: "Effects of Nonuniformities of Fiber Distribution on Thermally Induced Stresses and Cracking in Ceramic Matrix Composites", **Mechanics of Materials**, 1993, Vol.16, PP.351-363.

(7) B.F. Sorensen and R. Talreja: "Effects of Interphase and Coating on Thermally Induced Damage in Ceramic Matrix Composites", **High Temperature Ceramic Matrix Composites**, R. Nasalin, et al., eds., 1993, Woodhead Publishing Ltd., Cambridge, U.K., PP.591-598

(8) Pascal Louis and A.M. Gokhale, "Application of Image Analysis for Characterization of Spatial Arrangement of Features in Microstructures", **Metall. and Mater. Trans-A**, 1995, Vol. 26A, PP. 1449-1456.

(9) Vidas User Manual.

(10) K.H. Hanish and D. Stoyan, "Stereological Estimation of the Radial Distribution Function of the Centers of Hard Spheres", **Journal of Microscopy**, 1980, Vol. 122, Part II, PP. 131-141.

(11) Hanish, K.H., (1983), "On the Stereological Estimation of the Second Order Characteristics of Hard Core Spheres", Biometrics Journal, Vol. 25, PP. 731-738.

(12) M. Tanemura, "On the Stereology of Radial Distribution Function of Hard Sphere Systems", **Science of Form: Proceedings of First International Symposium on Science of Form**, Y. Kato, R. Takaki, and J. Toriwaki, eds., KTK Scientific Publishers, Tokyo, Japan, 1986, PP. 157-165.

(13) S. Chandrasekhar: "Stochaistic Problems in Physics and Astronomy", **Rev. Mod. Phys.**, 1943, Vol. 15, PP. 86-89.

(14) A.J. Ardell and P.P. Bansal: "Average Nearest Neighbor Distance Between Uniformly Distributed Finite Particles", **Metallography**, 1972, Vol. 5, PP. 97-111.

(15) H. Schwartz and H.E. Exner: "The Characterization of Arrangement of Feature Centroids in Planes and Volumes", **Journal of Microscopy**, 1983, Vol.129, PP. 155-169.

(16) B.D. Ripley: **Spatial Statistics**, John Wiley and Sons, London, U.K., 1981.

(17) B.D. Ripley: "Modelling Spatial Patterns", **J. Royal Stat. Soc.-B**, 1977, Vol. 39, PP. 172-212.

(18) S. Yang and A.M. Gokhale, Private Research, Georgia Institute of Technology, Atlanta, Georgia, 1995.

The Continuing Evolution of the SEM

Mel D. Ball

Alcan International Ltd., Kingston, Ontario, Canada

Abstract

Since its commercial introduction in 1965, the SEM has become an indispensable tool in many fields of research and technology. The basic imaging capabilities of the SEM still account for the majority of its uses. However, additional facilities and capabilities have been introduced and have extended the range of application significantly. A modern, fully equipped SEM is now capable of providing detailed information on the chemistry and crystallography of a wide variety of materials, as well as offering a versatile range of imaging possibilities.

Introduction

Fundamentally, scanning electron microscopy involves generating a fine beam of electrons, scanning this beam over the surface of a specimen, and collecting the resulting signals of interest for processing and display. The variety of techniques and detectors which have been developed to exploit the interactions between the electron beam and the specimen have resulted in specialised instruments. Figure 1 is a schematic illustration of the relationships between many of these modern analytical instruments and methods, and the basic SEM.

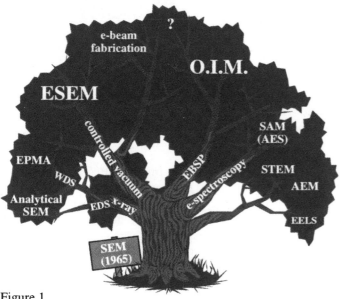

Figure 1

After a brief review of technical developments in scanning electron microscopy, some recent advances are described in more detail. The capabilities of the SEM for microanalytical characterisation and for crystallographic studies will be outlined. Of particular interest in this respect is "Orientation Imaging Microscopy", which offers significant opportunities in the study and characterization of crystalline materials. Another rapidly developing technology is "Environmental" scanning electron microscopy (ESEM), which has greatly extended the range of materials and processes which can be studied. Finally, some areas for future developments in Scanning Electron Microscopy will be considered.

Although closely related, instruments such as the Scanning Auger Microprobe (SAM) and the Scanning Transmission Electron Microscope (STEM) are generally regarded as distinct specializations and will not be discussed here. A number of specialised SEM techniques such as EBIC and imaging of magnetic domains have also been omitted.

Brief Review of SEM development

The original scanning electron microscopes offered a limited range of capabilities and had modest resolution (~ 20 nm). Nevertheless, the machines soon found wide ranging applications in both the physical and life sciences. Although they had a higher resolution than light microscopes, probably the greatest advantage that these instruments offered, was the large depth of focus. This enabled "rough" surfaces to be observed and studied in detail.

Incremental improvements in gun design, lens and polepiece design, electronics, detectors and vacuum systems have all produced improvements in resolution. To achieve the highest image resolution, high brightness electron sources (LaB_6 or Field Emission Gun (FEG)) are used, although for many applications, tungsten filament sources give acceptable results and are still widely used. The advent of energy dispersive X-ray spectrometers resulted in a widespread incorporation of X-ray microanalysis capabilities and this led to a significant expansion in the use of SEMs. The development and use of channeling patterns and contrast effects for crystallographic studies has also expanded the applications of the SEM in some fields of materials science. Many of these advances have exploited the dramatic improvements in computer technology which have occurred over recent years.

Table I lists some of the more important advances which have expanded the use of the scanning electron microscope and resulted in the highly versatile instruments of today.

Note that the dates given in this table are estimates of when a particular technique or function came into general use and do not necessarily correspond to initial development and early prototype introductions.

Table I: Approximate Time Line for Commercial Implementation

1960 - 70
- Microprobe Analysis (EPMA) – (X-ray wavelength spectrometers).
- First Commercial SEMs are introduced (1965) and find wide application.
- Resolution (SEI) ~ 15 - 20 nm

1970 - 80
- X-ray EDS introduced and widely adopted
- Backscattered electron detector development
- High resolution ("immersion lens") concept introduced.
- Higher brightness electron emitters (FEG, LaB_6)
- Microprocessor control of some functions is introduced.
- Image analysis - digital scan generation is introduced.
- Standard SEI resolution ~ 7 nm
- High resolution (immersion) ~ 3 nm
- High resolution (with FEG) ~ 2 nm

1980 - 90
- EBSP systems introduced.
- Microprocessor controlled electron optics.
- Improved gun and column design for Low Voltage SEM
- UTW detectors for X-ray microanalysis of low Z elements.
- Introduction (in 1988) of first commercial ESEM.
- High resolution (SEI) ~ 1 nm
  ~ < 5 nm at 1 kV

1990 -
- Integrated digital scanning and image processing.
- Orientation Imaging Microscopy (OIM) is introduced.
- Resolution ~ 3 nm at 1 kV.

Compositional Information and Microanalysis

When a beam of electrons impinges on a sample, some of the interactions which occur are composition dependant and with suitable detectors, the SEM can be readily adapted for microanalytical studies at high spatial resolution. In the SEM, the electron beam can be focussed to a spot size of a few nanometers (or less) and precisely located on a feature of interest. However, in practice, scattering of the electron beam within the sample generally means that the effective resolution (the region from which the information is generated) is of the order of 1μm (although in many cases, much better spatial resolution can be achieved by careful selection and optimisation of the operating conditions for the particular investigation).

The emission of X-rays with energies (and wavelengths) which are characteristic of the elements present in the irradiated region, forms the basis for X-ray microanalysis and this is probably the most widely used analytical method in the SEM. Electron backscattering is also composition dependant and can be useful for characterising suitable samples.

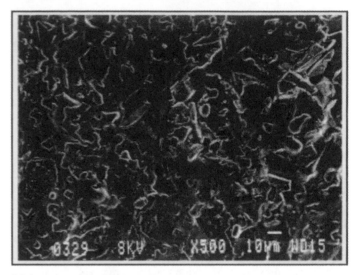
(a)

(b)

Fig 2 – Surface of an aluminum-based composite material
a) Secondary electron image; b) Backscattered electron image (same area). Atomic number contrast reveals the presence of several phases.

Backscattered Electron Imaging. Backscattered electrons result from high angle scattering by atomic nuclei within the specimen (Mott scattering). Higher atomic number elements have larger scattering cross-sections so that the intensity of the backscattered electron signal will be related to the atomic numbers of the elements present in the region of interest. By incorporating high performance backscattered electron detectors into the SEM, this atomic number contrast mechanism provides some compositional information about the specimens. Figure 2, shows

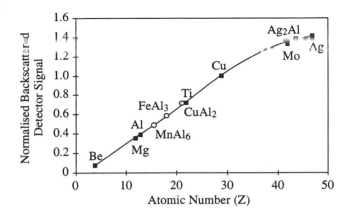

Fig 3 – Variation of the Backscattered Electron Signal with atomic number for a JEOL 840 SEM at 10kV (solid state backscattered detectors).

the comparison between a secondary electron (SE) image of the surface of an aluminum based composite material and the corresponding backscattered electron (BE) image. In the BE image, the atomic number (Z) contrast clearly reveals the presence of at least three distinct phases which are difficult to distinguish in the secondary electron (SE) image. In this example, the dark (low atomic number) particles are silicon carbide, the intermediate (grey) contrast is the aluminum matrix, and the brightest (highest Z) particles are an Al-Fe-Si intermetallic phase. For suitable specimens (eg metallographically polished), quantitative atomic number measurements can be obtained from backscattered images. To do this, the variation of the backscattered signal (grey level) with specimen atomic number for the SEM / detector system must be established. As an example, Figure 3 shows the backscattered electron signal intensities

which were recorded from standard samples of known atomic number. The measurements are normalised in this case with respect to the signal from a copper standard (Z=29). Using this calibration relationship, the backscattered signal from any unknown phase can be compared to the signal from a standard (eg Cu) and the apparent atomic number can be determined. This ability to relate image grey levels to compositional information can be especially powerful when used with image analysis applications, since it can often enable different constituent phases to be distinguished.

X-Ray Microanalysis and Mapping. Core level ionisation of atoms in the specimen by the electron beam, leads to the generation of characteristic X-rays. By collecting and counting these emitted X-rays and processing the data, the chemical composition of the region of interest can be deduced. The volume of the sample analysed will depend on the incident beam energy (keV) and the specimen density. Specialized electron probe microanalysers (EPMA) have been developed in order to fully exploit this phenomenon and can generate quantitative compositional analysis data from carefully prepared samples. Modern EPMAs are closely related to SEMs and incorporate computer controlled X-ray wavelength dispersive spectrometers (WDS), automated specimen stages and light microscopy as well as the secondary and backscattered electron detectors. For many applications, microanalysis systems based on X-ray energy dispersive spectrometers (EDS) can provide adequate compositional information and these systems are widely used with standard SEMs. When they were originally introduced, the thin beryllium window on the detector prevented the detection of X-rays from the light elements (below Z~10). However, the development of Ultra Thin Window (UTW) detectors now enables all elements of Z> ~5 (boron) to be detected. Improvements in detector (energy) resolution, collection efficiency and spectrum processing have also been made and this enables quantitative analysis of suitable samples to be obtained with reasonable precision. It is often useful to use the X-ray data to generate elemental distribution maps of particular regions of interest. In Figure 4, the backscattered electron image of an anodised aluminum surface is displayed along with X-ray maps showing the distribution of aluminum, oxygen and nickel. The sample had been strained to crack the oxide film so as to reveal the underlying aluminum metal surface. The nickel had been electrodeposited in the porous anodic oxide film to produce a black finish and the Ni K X-ray map shows that it is concentrated near the bottom of the oxide layer.

Crystallography in the SEM

Techniques such as X-ray diffraction and neutron diffraction are widely used for determining crystal structures and orientations. However, for fine grain polycrystalline materials diffraction data cannot be obtained from individual grains using these methods. In principle, electron diffraction in the transmission electron microscope can provide diffraction information from regions below 1μm in size. However, the diffraction of electrons by crystalline materials can also be exploited in the SEM and in many cases, this approach has advantages over the alternatives.

(a) (b)

(c) (d)

Fig 4 – X-ray distribution maps for Al, O and Ni for an anodised aluminum sample; a) Backscattered Electron Image; b) Oxygen (K_α) map; c) Aluminum (K_α) map; d) Nickel (K_α) map.

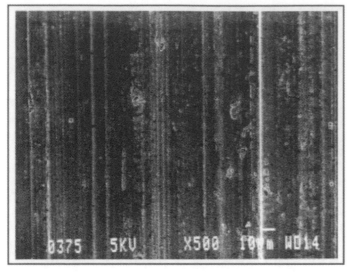

(a)

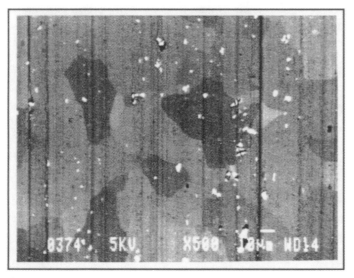

(b)

Fig 5 – Surface of a rolled aluminum sheet;
a) Secondary electron image showing the roll lines; b) Channelling Contrast Image (Backscattered Electron Detector) showing the grain structure.

Channeling . When an electron beam impinges on the surface of a crystalline material, the average penetration distance of the electrons will vary depending on the particular orientation and nature of the crystal lattice at that location. For some orientations, the electrons will interact strongly near the surface and give rise to a relatively high backscattered electron signal. However, for certain other orientations, the electrons will, on average, penetrate further into the sample before being scattered and they will have a lower probability of escaping from the specimen. For these "channeling" orientations, the backscattered electron signal will be lower. For polycrystalline materials, the backscattered signal will vary from grain to grain depending on the particular orientation, and a channeling contrast image will

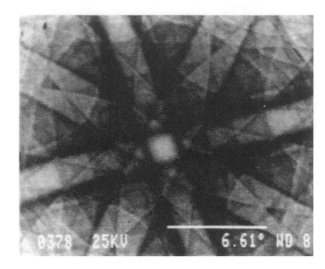

Fig 6 – Selected Area Channelling Pattern of a {100} oriented grain.

result. In Figure 5, a secondary electron image of the surface of a rolled aluminum sheet sample (chemically etched) shows the rolling lines and surface topography and the backscattered electron image reveals the channeling contrast from the aluminum grains. In some cases the grain boundaries are associated with the rolling lines, indicating that the grain growth process was influenced by the surface topography.

The Selected Area Channeling Pattern (SACP) technique uses the same phenomenon to determine the crystallographic orientation of small regions only a few microns in diameter. To produce these patterns, the electron beam is "rocked" at the location of interest on the sample so that the orientation of the beam relative to the crystal structure is varied systematically. As the beam orientation changes relative to the crystal lattice, the backscattered electron signal varies accordingly to produce the SACP. As an example, Figure 6 is a SACP obtained from a FCC crystal in the cube {001} orientation.

Electron Backscattered Patterns (EBSP). The use of Electron Backscattered Patterns (also referred to as Backscattered Kikuchi Patterns) to study crystalline materials is a more recent development and provides a method of studying the crystallography of sub-micron regions. In this technique, the specimen is tilted from normal incidence by a known angle (typical 70 degrees). The incident electron beam is positioned on the region of interest and scattering within the specimen produces backscattered electrons. Since these backscattered electrons typically originate from a small subsurface region, they will interact with the crystal lattice as they emerge through the surface layer. The diffraction and scattering which occurs in this region results in variations in intensity in the backscattered electrons. This distribution can be recorded using a phosphor screen and a sensitive video camera to produce the EBSP. An example of an EBSP obtained from a small grain is shown in Figure 7. Because of the large collection angle, these patterns are relatively easy to index (see Figure 8) and, provided the geometry of the sample and screen is known, the orientation of the grain can be determined.

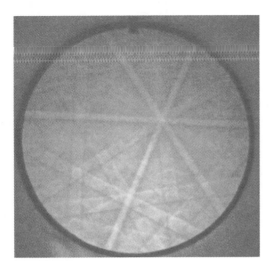

Fig 7 – An EBSP from a small grain in a recrystallised aluminum sheet sample.

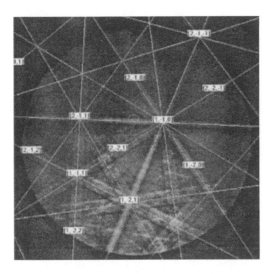

Fig 8 – The same EBSP with the principal bands plotted and indexed.

Orientation Imaging Microscopy . By using a computer workstation to automate the collection and processing of electron backscattered patterns, it is now possible to generate images which are based on the orientation variations within the area of interest. The electron beam is moved to a predetermined array of positions on the specimen and the EBSP is recorded at each of the locations. Each pattern is processed by the computer to determine the orientation of the crystal lattice at that location. By evaluating the "quality" of the pattern, an estimate of the degree of deformation of the material at that location can also be provided. The results can be displayed in a variety of ways and some examples are given in Figures 9—12. The data set was obtained from a 1mm square region of a recrystallised aluminum sample and is based on the collection and analysis of 40,000 patterns (in a 200 x 200 array).

Figure 9 is a map of the pattern "quality" variation within the region of interest. The brighter pixels correspond to regions of low lattice deformation which produce clear, good quality

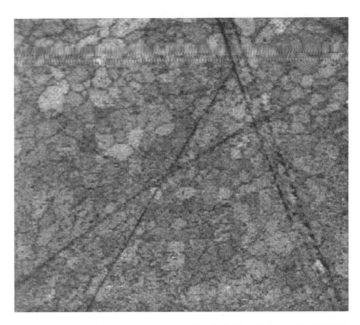

Fig 9 – A map of the "pattern quality" derived from the EBSPs from a 200 x 200 point array. Strain associated with scratches from sample preparation is detected as dark lines.

patterns, while the darker regions correspond to the locations of patterns which were poorly defined. In some cases, these darker pixels coincide with grain boundaries and in other cases they may be due to the presence of intermetallic particles or plastically deformed regions. In Figure 10 all the positions where orientation changes of greater than 15 degrees occured have been identified as grain boundaries (black lines). A few small angle boundaries (with misorientations between 5 and 15 degrees) have also been outlined (thin black lines).

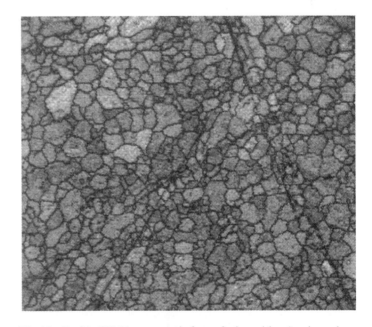

Fig 10 – In this OIM image, grain boundaries with misorientations greater 15° have been highlighted (black lines). A few low misorientation boundaries were detected and are represented by thin black lines.

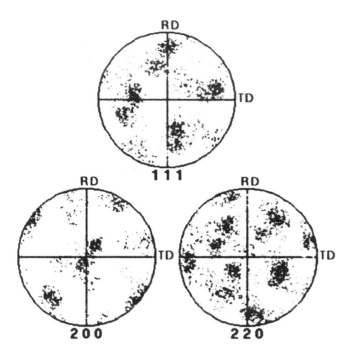

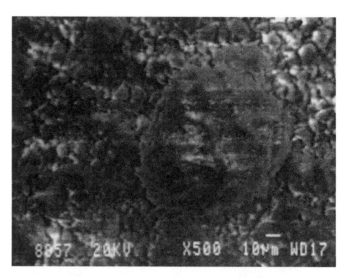

Fig 11 – Pole figure plots for the same area. The orientation of each pixel is plotted.

Specimen Considerations

Clean, dry, conductive materials are readily studied in the SEM with little or no specimen preparation. However, for other materials, problems can arise due to specimen charging or electron beam damage. In many cases these problems can be minimised or eliminated by choosing appropriate imaging conditions or by coating the specimen with a thin conductive film (e.g. sputtered gold or carbon). Figure 13 shows a sample of alumina powder which becomes charged during SEM observation at 20 kV causing a substantial degradation of the image. This problem is virtually eliminated if the SEM is operated with a much lower electron beam energy (3kV in this example). At this lower kV, the secondary electrons are generated closer to

In Figure 11, the orientation of each indexed pattern from selected grains has been plotted on standard pole figures to show the crystallographic texture components. In Figure 12, all the grains with a particular range of orientation have been highlighted. In this sample and for the orientation selected, the grains appear to be randomly distributed. In other samples, clustering of grains with similar orientations has been observed.

a)

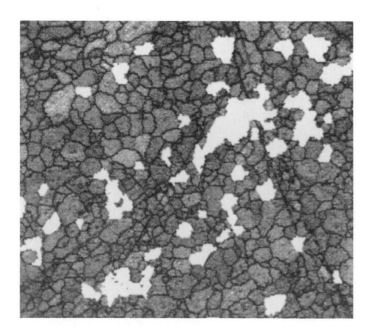

b)

Fig 12 – In this image, all the grains close to a particular orientation have been highlighted white.

Fig 13 – SEM images of uncoated alumina powder particle; a) 20 kV image showing severe charging and image degradation; b) 3 kV image of the same particle shows no noticeable charging problems.

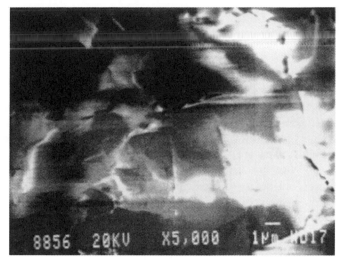

a)

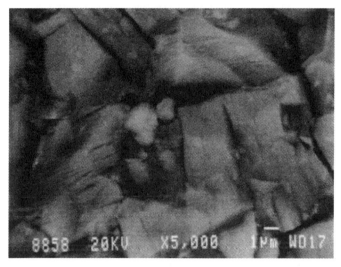

b)

Fig 14 – Another example of image degradation due to charging for an alumina powder particle;
a) Secondary Electron Image at 20 kV; b) Backscattered Electron image of the same region is not significantly affected by charging.

the specimen surface and the secondary emission coefficient is enhanced so that very little charge accumulates. An alternative approach to the study of insulating specimens is exemplified in Figure 14. In this sample of an alumina (Al_2O_3) powder, the secondary electron image is severely degraded due to the build up of charge. However, the backscattered electron image (also recorded at 20kV for comparison) provides a satisfactory result. In this case, the high energy backscattered electrons are not significantly affected by charge accumulated on the specimen. The recently introduced Environmental Scanning Electron Microscopes also eliminates charging problems and can enable an even wider range of materials to be studied under controlled conditions.

The Environmental Scanning Electron Microscope (ESEM)

The ESEM is designed to allow investigations to be carried out with a controlled pressure of a selected gas or vapour in the specimen region. The successful operation of this instrument relies on the fact that a significant fraction of the beam of electrons will pass through a small region of low pressure gas without being scattered. By maintaining a high vacuum in the gun and column, the scattering of the electron beam will only occur in the specimen chamber and will depend on the electron beam energy (operating voltage), the gas pressure and the working distance (from the final aperture to the specimen).

To achieve the desired performance, a number of significant technical developments were necessary. A carefully designed and controlled differential pumping system was required to maintain the high vacuum in the electron gun and upper column and, at the same time, to allow the specimen chamber pressure to be independently controlled. To maintain the pressure differences, small apertures are incorporated to restrict the leakage of gas from the specimen chamber into the upper column, where it can be efficiently pumped away. Water vapour, air, nitrogen or other gases can be used in the specimen chamber at pressures ranging from < 1Pa to ~ 2.5 kPa (1 torr ~ 133 Pa). The selected gas or vapour is supplied to the specimen chamber through controlled needle valves. Note that in a conventional SEM, the specimen chamber pressure would be ~ 10^{-3} Pa or less.

Since the Everhart - Thornley secondary electron detectors used in standard SEMs can only operate in a good vacuum, an alternative detector system had to be developed for the ESEM. The ESEM detector (or GDD detector) utilises a biased grid electrode above the specimen. Electrons emitted from the specimen surface are accelerated by the applied field and cause ionisation of the gas. The ionisation current is detected by the grid and is used as the image signal. Backscattered electron detectors and X-ray (EDS) detectors can also be used with the ESEM.

Fig 15 – Polymer particles on an anodised aluminum oxide surface using an Electroscan ESEM (20 kV, 10 mm working distance, chamber pressure ~200 Pa [air]).

By choosing suitable operating environments, insulating specimens, including ceramics and polymers, can be observed with no charging problems because ionised gas molecules near the specimen surface quickly neutralise any charge build up. Generally, relatively low gas pressures can be used where the main concern is the elimination of charging. For example, in Figure 15, a dispersion of polymer particles on an insulating aluminum oxide surface could be observed at 20kV with a chamber pressure of less than 200 Pa of air and a working distance of 10mm. Figure 16 shows a thin aluminum oxide film which is partly adhered to a plastic film substrate. In this case, a pressure of 50Pa (air) was sufficient to eliminate specimen charging (also at 20kV and 10mm working distance).

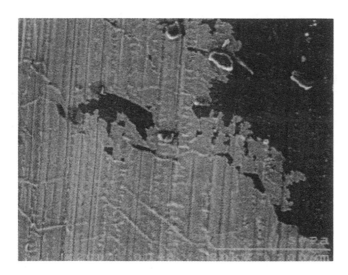

Fig 16 – Thin aluminum oxide film on a plastic substrate (20 kV, 10 mm working distance, chamber pressure ~50 Pa [air]).

The ability to control the local environment around the sample also allows wet or oily specimens to be examined. Furthermore, the incorporation of special stages for heating and cooling of the specimen enables dynamic in situ experiments to be carried out. For example, with this instrument it is now possible to observe moisture condensation or evaporation from surfaces. The growing range of applications includes fields such as corrosion, lubrication, adhesives and polymer coatings in addition to many areas within the life sciences.

While the opportunities for the ESEM are clear, the novel features of the technique and particularly the effects of electron scattering by the gas, need to be considered carefully when interpreting the results. This can be important in the case of X-ray microanalysis of small features. Although the central, unscattered part of the electron beam will generate data from the targeted region of interest in the normal way, the electrons scattered by the gas molecules will be spread over a much wider region and X-rays generated from this extended area will contribute to the overall result. The effect of this contribution will depend on the beam energy (kV), gas pressure and working distance. The scattering of the electron beam can be minimised by using a relatively high operating voltage, and the lowest useful working distance and pressure.

The Future of the SEM

Driven by the need for improved material properties and performance, the applications for the SEM in materials research and development continue to grow. The SEM is also becoming a valuable tool for quality control in a wide range of industries. With the continuing development of the ESEM, the range of materials which can be studied is growing. Furthermore, the possibilities of carrying out in-situ studies without the limitations of high vacuum is also extending the role of the ESEM.

With a modern SEM, it is already possible to automate the specimen stage movement, control the electron beam and acquire and process digital images. In principle, a suitable computer workstation could also be interfaced to an energy dispersive spectrometer system and the camera of an EBSP system so that chemical and crystallographic information can be generated. With suitable image processing and analysis software, a comprehensive specimen characterisation would be possible. This fully integrated system concept is outlined in Figure 17. After an initial setup for the specimen, such a system would be capable of carrying out sophisticated image analysis, chemical microanalysis and/or crystallographic characterisation of a predetermined series of fields of view.

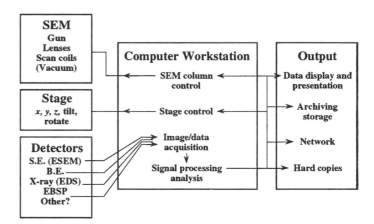

Fig 17 – Schematic diagram of a fully inegrated SEM system with image analysis, X-ray microanalysis and EBSP capabilities.

Acknowledgments

Paul Nolan, Queen's University, Kingston, Ontario – for assistance with the ESEM.

David Dingley, TexSEM Laboratories Inc. Provo, Utah – for providing OIM micrographs and data.

Alcan International Limited, Kingston, Ontario – for permission to present and publish this work.

References

J.I. Goldstein, H.Yakowitz, D.E.Newbury, E.Lifshin, J.W.Colby and J.R. Coleman, "Practical Scanning Electron Microscopy", Plenum Press, New York and London (1975).

Joy, D.C., Mat.Res.Soc.Symp.Proc. 332, Materials Research Society (1994).
"The Field Emission Gun Scanning Electron Microscope - High Resolution at Low Beam Energies"

Joy, D.C., D.E.Newbury,D.L.Davidson, J.Appl. Phys., 53, R81-R122, (1982)
"Electron Channeling Patterns in the Scanning Electron Microscope"

Venables, J.A. and C.J. Harland, Phil. Mag.27, 1193-1200 (1973)
"Electron Backscattered Patterns - a new technique for obtaining crystallographic information in the SEM".

Dingley, D.J., Scanning Electron Microscopy 1984/II, SEM Inc., (Chicago) 569-575.
"Diffraction from submicron areas using electron backscattering in a SEM".

Mason, T.A. and B.L. Adams, JOM., 46, 43-45 (1994)
"The application of Orientation Imaging Microscopy".

Adams, B.L., S.I.Wright and K.Kunze, Met.Trans. 24A, 819-831 (1993)
"Orientation Imaging: The emergence of a New Microscopy"

Wright, S.I., B.L.Adams and K.Kunze, Mat. Sci. Eng. A160, 229-240 (1993)
"Application of a new automatic lattice orientation measurement technique to polycrystalline aluminum"

Danilatos, D.G., J.Microscopy 160, 9-19 (1990)
"Mechanisms of detection and imaging in the ESEM"

Danilatos, D.G., Mikrochim.Acta, 114/115, 143-155 (1994)
"Environmental Scanning Electron Microscopy and Microanalysis"

Cameron, R.E. and A.M.Donald, J.Microscopy 173, 227-237, (1994)
"Minimizing sample evaporation in the environmental scanning electron microscope"

Developments in Energy-Dispersive Spectroscopy

John J. Friel
Princeton Gamma-Tech, Princeton, NJ

Abstract

Energy-dispersive spectroscopy has been available as a technique for X-ray microanalysis since the late 1960s. But it was not until the 1970s that it became a standard accessory on SEMs. Detectors can be made from lithium-drifted silicon or from germanium, but Si(Li) is predominant. Developments in EDS have been made both in the detector with its associated electronics, and in the computer analyzer and its software. Detector resolutions have improved and throughput has never been greater. There exists a choice of windows, and light element analysis has become commonplace. Quantitative analysis routines have been optimized, and computer-aided imaging is routinely performed on the EDS computer. Even X-ray maps that were once only used to document the spatial distribution of elemental concentration now can be processed by the computer to reveal information that was heretofore inaccessible.

Energy-Dispersive Spectroscopy is a widely used tool for microanalysis on both scanning electron microscopes (SEMs) and electron probe microanalyzers (EPMAs). In a discussion of developments in any field, it is usually the recent ones that people find most interesting. For this reason, emphases will be placed on developments within the last ten years or so, but not to the exclusion of the early discoveries and continuing advances over the years. A discussion of energy-dispersive spectroscopy (EDS) logically divides into two lines of development: X-ray detector/hardware and computer analyzer/software advances. This paper will follow these two aspects in parallel, citing those innovations that in the opinion of the author were the most important and the most widely accepted. Although it is possible to build EDS detectors out of several materials, this discussion will be limited to solid-state devices made out of lithium-drifted silicon (Si(Li)) or high-purity germanium. These are by far the most common X-ray detectors on electron microscopes.

The 1960s

The application of a solid-state Si(Li) detector on an electron column instrument was first described by Fitzgerald, et al. (1) on an electron probe microanalyzer. Its active area was 50 mm^2, and it had a resolution of about 700 eV at manganese. The beryllium entrance window was 0.125 mm thick and effectively blocked the transmission of all elements lighter than calcium ($K\alpha = 3.7$ eV) This type of detector was considered quite an advance, because it collected an entire spectrum at once without having to scan spectrometers, and it was also insensitive to beam scanning that causes defocusing in crystal spectrometers.

Interestingly, Fitzgerald et al. (1) realized that the high collection efficiency of about 80% at the time permitted analysis at lower beam current for the same counting statistics and electron dose on the specimen. As we will see, later in the 1990s this advantage was more fully exploited with digital pulse processing.

The first installation of a Si(Li) detector on an SEM came in 1968 (2). The marriage of solid-state detectors and SEMs is natural, because the small probe size necessary for high resolution imaging produces low X-ray count rates. At this time spectra were recorded on a multi-channel analyzer (MCA), without internal calibration. The instrument had to be calibrated to the analyzer instead of the other way around. The output was merely the number of counts in each channel, and software to manipulate or analyze the spectrum did not exist.

The 1970s

Detector

During the 1970s, the number of SEMs with EDS detectors grew rapidly, as those who bought the microscope for its imaging versatility and ease of specimen preparation learned the value of microanalysis. Metallurgists found that they could get the actual composition of inclusions and small second-phase particles in the SEM with no more specimen preparation than required for the light microscope. The spectral resolution of the detectors improved over time to under 160 eV. Figure 1a shows a spectrum of an alloy of AgPd collected in 1976. The elements are adjacent in the periodic table and their Lα peaks are separated by 146 eV. The spectra in the Figure 1b show an expanded peak envelope and spectra of pure Ag and pure Pd beneath it. It is evident that at this time spectral resolution was inadequate to separate these peaks, though deconvolution software made quantitative analysis still possible, as we will see. Also note in Figure 1 the crude displays of the time.

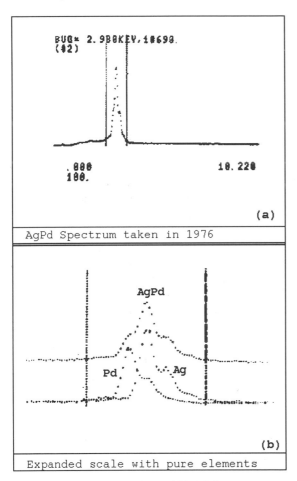

Fig. 1 Spectra of AgPd alloy from 1970s EDS system

Light-element Detectors. Light element performance was also an issue. The beryllium window thickness was reduced to about 7 μm—the thinnest sheet that can be rolled without pinholes. This made it possible to analyze sodium and all elements heavier. And after Jaklevic and Goulding (3) built an experimental ultra-thin window (UTW) detector, commercial windowless detectors became available (4). Actually these were multiple window detectors, with beryllium foil that would withstand atmospheric pressure in one position and no window or an ultra-thin window in another. The advantage of a windowless detector, of course, was that X-ray analysis on an element as light as carbon was possible, but the disadvantage was that the Si(Li) crystal was exposed to the lower vacuum of the microscope. Nevertheless, this type of window arrangement

was used for light element analysis throughout the 1980s, although beryllium window detectors were still the standard.

EDS on Microprobes. Though the SEM was the most common instrument using EDS, the technique also gained favor on EPMAs. The EDS system on a probe produced a quick qualitative analysis, and it was common to do an EDS survey before a quantitative wavelength-dispersive (WDS) analysis.

EDS on TEMs. Even transmission electron microscopes (TEMs) were fitted with EDS detectors. Again the high efficiency of the solid-state device was useful for analyzing very small particles and even biological specimens. With the advent of a focused probe and scanned beam, the scanning transmission electron microscope (STEM) opened new frontiers in small volume, high sensitivity analysis. Those instruments with EDS and possibly electron energy loss spectrometers (EELS) are called analytical electron microscopes (AEMs).

Analyzer

The most important advances in this decade probably were made in software and in the computers that ran it. Some manufacturers chose to collect a spectrum directly into computer memory rather than into an MCA. This approach eliminated memory halves and the fixed eV/channel. Software became available to manipulate spectra and to compare more than two spectra, and the way a spectrum was displayed was no longer linked to the way it was collected.

Software for Quantitative Analysis. The principal advance in software, however, was in quantitative analysis. Heretofore, this was restricted to the microprobe. But in 1976, the National Bureau of Standards released the computer program, Frame C (5). It implemented the ZAF procedure using the equations of Philibert, Duncumb, and Heinrich for the atomic number (Z) and absorption correction (A) and that of Reed for the fluorescence correction (F). Earlier versions of the Frame program were meant for EPMA data, but Frame C incorporated a background modelling and subtraction routine as well as a peak deconvolution routine by the method of overlap factors. Both of these are necessary for EDS because of its nonlinear background and poorer resolution than WDS.

Separately from NBS and before publication of Frame C, an alternative method of handling the background and peak overlaps was proposed (6). In 1973, Schamber suggested using a digital filter for background removal and using multiple-least squares fitting for peak deconvolution. The digital filter merely treated the background as a slowly changing waveform, i.e., a low-frequency signal, and filtered it out. This method had the advantage of handling all spectra empirically and not assuming any knowledge of the actual spectrum.

The computers of the day were minicomputers, and they had the advantage of allowing instruments to have their own dedicated computer. Prior to these, computing the matrix correction for quantitative analysis was often done as a batch job on a mainframe. But despite the advantage of having an entire computer devoted to one instrument, the minicomputers had limited memory, and mass storage at first was magnetic or paper tape.

The only disadvantage to the approach of Schamber was that it took more computer time than the theoretical approach of Frame C. Later, as computers gained power, most EDS manufacturers began to use the filter and fit method.

Standardless analysis programs were also developed in which the intensity ratio was calculated from a theoretical pure element standard. These work well for some materials and less well for others. Alloy steels such as the various stainless steels work particularly well, probably because the fundamental constants do not vary greatly among the transition metals.

Combined EDS/WDS. As previously mentioned, EDS systems were being installed on microprobes, so it made sense to run the quantitative software for both WDS and EDS in one computer. In fact, automation of the wavelength spectrometers and stage could also be controlled by the computer. The first truly combined EDS/WDS analysis was done in 1977 (7). In this analysis, an entire spectrum was collected by the EDS system and 14 elements were collected sequentially by WDS, including all the rare-earth elements. The data were combined at the matrix correction step. The data for a total of 213 elements took 12 mins. to collect and another 25 minutes of minicomputer time to perform the ZAF matrix correction.

By the end of the decade, EDS was a routine method of microanalysis available on most SEMs. Many thought that the technique was mature both in detector specifications and in analytical algorithms and software to implement them. Such proved not to be the case, as considerable advances remained to be made.

The 1980s

Despite the complacency of the late 1970s, the 1980s saw the invention of a new window type, use of germanium detectors, and greatly improved spectral resolution. On the analyzer side, we witnessed much development in matrix correction programs of the phi-rho-z type, leading to many commercial products. We went from minicomputers to microcomputers using large-scale integrated circuits, all the way to networked workstations by the end of the decade.

Detector

Germanium Crystals. Experiments with germanium detectors spanned the decade. Because it takes less energy to produce an electron-hole pair in germanium, it has better intrinsic resolution than does silicon. It also has higher efficiency for high-energy X rays (see Figure 2). In a paper in 1981, Barbi and Lister compared the two materials and concluded that germanium was unsuited for the analysis of X-ray lines below about 2 keV, owing to incomplete charge collection (8). This manifests itself as a long tail on the low energy side of the peak (see Figure 3). At that time germanium was considered unsuitable for light-element detectors. Work continued on the material, however, and a later update comparing the two materials was published by Cox et al. (9).

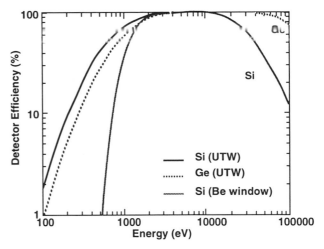

Fig. 2 Diagram of Ge and Si Efficiency

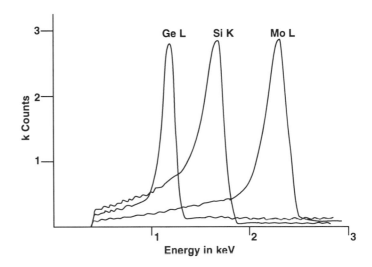

Fig. 3 Low Energy Peak Shape — IG Detector

Atmospheric thin windows. The most significant innovation of this time was no doubt the atmospheric thin window detector (ATW) that could withstand the differential pressure between the vacuum of the detector and atmospheric pressure when the SEM chamber was open (10). Up until this time, beryllium windows were standard, and some multi-position windowless detectors were used. Although any window absorbs more than does a windowless detector, most people are willing to accept that drawback for the convenience of not having to worry about contaminating the detecting crystal with ice or oil or forgetting to shut the window before venting the microscope.

The original atmospheric thin windows were made out of an engineered material consisting of boron and nitrogen. Later, other manufacturers invented their own or bought them from an outside source. The field seems to have settled on polymers as materials that can be made thin and still be strong and tough. Polymers also absorb carbon photons very little; thus that important element is easily analyzed through such a window.

The market seems to have switched almost exclusively to this type of detector. Some beryllium window EDS detectors are used on EPMAs, because light elements can be analyzed with the crystal spectrometers. And some windowless detectors are still used for the ultimate in analytical sensitivity. With an atmospheric thin window detector, it is possible to detect light elements down to and including beryllium. Boron is easily detected, and carbon and oxygen are routine.

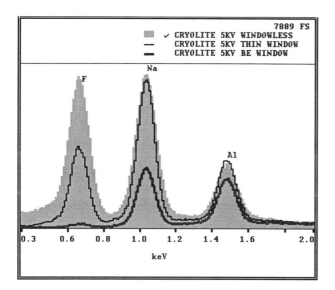

Fig. 4 Spectra comparing different windows

Spectra of the mineral cryolite containing Na, Al, and F that were taken in 1988 are shown in Figure 4. This figure compares spectra from a turret type detector fitted with a beryllium window, a windowless position, and an atmospheric thin window. Note that there is some loss of sensitivity at fluorine for the thin window when compared with the windowless detector, but they have about equal sensitivity at sodium. In all cases, the Be-window detector is less sensitive even up to aluminum, though they are all about equally sensitive at about silicon and above.

Analyzer

Quantitative Analysis Algorithms. A phi-rho-z curve is a plot of the number of ionizations in a material as a function of mass-depth. If one knows the depth distribution of X rays generated and the same distribution after some are absorbed in the material, then one can calculate the atomic number and absorption factors. The fluorescence factor is calculated separately as in ZAF. The concept of a phi-rho-z curve goes back to the first microprobe, but it was not until a paper by Packwood and Brown in 1981 (11) that the concept of using calculated curves to perform a matrix correction got started.

Following their paper, groups in various countries started calculating such curves and writing phi-rho-z programs, mainly for WDS data, but they work for EDS as well. Although there are too many articles on the subject to cite, the principal researchers include Packwood and Brown in Canada, Love and Scott in the UK, Pouchou and Pichoir in France and Bastin in the Netherlands. Armstrong in the U.S. has also produced programs that give one a choice of algorithm. All of these groups claim superiority of their program, especially for light element analysis.

The phi-rho-z method shows a small improvement over the traditional ZAF method in analysis of metals having one or more elements that are significantly lighter or heavier than the rest. Analysis of B, C, and O would certainly be improved. But for routine microanalysis of metals, there is little difference, especially when one considers the precision of the analysis. Precision is dominated by counting statistics and is rarely limited by the matrix correction program.

A comparison of methods is shown below. This analysis is of a complex mechanical alloy containing eight elements. A wet chemistry analysis is shown to the right. In this alloy, aluminum is the lightest element, and one can see that the absorption correction only changes from 2.84 to 3.30. The phi-rho-z method produced a better result, but both are well within the range of the precision of the analysis.

Comparison of ZAF and phi-rho-z Methods

ZAF Method							
Element	k-Ratio	Z	A	F	ZAF	Wt%	Wet Chem
Al K	0.01320	0.9297	2.8395	0.9996	2.6387	3.48	4.00
Ti K	0.02310	1.0077	1.0663	0.9378	1.0077	2.33	2.26
Cr K	0.16190	1.0083	1.0357	0.9185	0.9592	15.53	14.76
Fe K	0.01070	1.0051	1.0430	0.8541	0.8954	0.96	1.00
Ni K	0.68650	0.9868	1.0233	0.9963	1.0060	69.06	68.12
Mo L	0.01450	1.1151	1.2994	0.9924	1.4379	2.08	1.91
Ta M	0.01360	1.1893	1.5030	0.9975	1.7830	2.42	1.91
W M	0.01900	1.1912	1.4660	0.9971	1.7412	3.31	3.91
TOTAL						99.18	97.87

phi-rho-z Method							
Element	k-Ratio	Z	A	F	ZAF	Wt%	Wet Chem
Al K	0.01320	0.9217	3.3013	0.9996	3.0415	4.01	4.00
Ti K	0.02310	1.0067	1.0625	0.9376	1.0029	2.32	2.26
Cr K	0.16190	1.0063	1.0332	0.9183	0.9547	15.46	14.76
Fe K	0.01070	1.0027	1.0406	0.8538	0.8909	0.95	1.00
Ni K	0.68650	0.9848	1.0225	0.9965	1.0034	69.89	68.12
Mo L	0.01450	1.0800	1.2886	0.9924	1.3811	2.00	1.91
Ta M	0.01360	1.1985	1.4187	0.9976	1.6962	2.31	1.91
W M	0.01900	1.1996	1.3774	0.9972	1.6477	3.13	3.91
TOTAL						99.07	97.87

Standardless Quantitative Analysis. In addition to improvements in the matrix correction, at this time standardless quantitative routines were developed. These came from the EDS manufacturers in response to the needs of SEM/EDS users. In an SEM, inaccuracies are introduced by a specimen change between standards and unknowns; precise and reproducible geometrical configuration is hard to achieve, and beam current drift may be a problem. These problems are all obviated with standardless analysis.

In the late 1980s, ASTM instituted an interlaboratory test program to evaluate quantitative EDS for metals using both standards and standardless methods. The results are published in a Standard Guide to Quantitative Analysis by Energy-Dispersive Spectroscopy (E 1508) (12). They found that, except in the case of elements present at less than about 10 wt%, standardless methods were as good as or better than using standards. Even though this result should not be surprising, because of the effects mentioned above, any standardless program should be thoroughly evaluated before placing faith in its results. Biases may exist in any program, and thus the method should be validated against standards before using it for any given class of materials.

Digital X-ray maps. Because knowing the distribution of elements in a specimen is often important, X-ray maps have been collected since the early days of electron probe microanalysis. Maps can be used to ascertain which elements associate with each other, and even to discern concentration gradients. Analog maps are sometimes called "dot maps" because they are constructed by putting dots on the screen whenever a photon is detected within a chosen energy region. Concentration of an element is inferred from clustering of the dots. In the 1980s, digital X-ray maps became available. In these, the actual number of X-ray counts is stored in computer memory, and represented on the screen by the intensity of the display. Digital maps do far more than display spatial information; they can be used to obtain information that would not otherwise be available without a computer to process the large dataset.

One can have the computer screen the individual elemental maps for ranges of concentration and even make Boolean comparisons among the elements. In this manner, specific phases can be extracted or quantitative comparisons among phases can be made. This capability is usually called image math or a similar name, and it is one of the most powerful techniques for understanding the spatial distribution of composition of a specimen. An example of such comparison is shown in Figure 5.

The backscattered electron photomicrograph in the upper left is of a ceramic superconductor taken in 1987 shortly after they were discovered. Although the only obvious second phase is some dark copper-rich unreacted starting material, the X-ray images show more phases. The backscattered electron image lacks the contrast to distinguish two phases that are very close in atomic number. Because the superconducting "123" phase differs greatly in composition from the "211" phase, they are easily distinguished in the X-ray images of Ba, Y, and Cu. The other images in the figure are the result of screening operations. The data in both the barium and yttrium images were screened by the computer for specified compositional ranges. Pixels that met the criteria in both images were then displayed on the bottom row. The operation was carried out twice with different compositional ranges for the 123 phase and separately for the 211 phase. Once these phases are extracted, measuring their area fraction is a simple image analysis operation. Inasmuch as the amount of the superconducting phase is crucial to

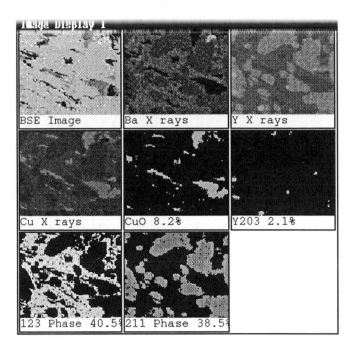

Fig. 5 Boolean operations on digital X-ray maps

the properties of the material, the image math operation is essential to making that measurement.

Workstations. The end of the 1980s saw some manufacturers move their analyzers from LSI 11 processors to desktop workstations. This brought the immediate advantage of better graphics and built-in networking. The workstations at the time were about four times faster at integer operations than the LSI processors they replaced. But speed was not the only advantage they possessed; it was not even the most important. The much improved graphics and multi-tasking capability allowed image analysis and image processing to be done on the same system as EDS. Some image analysis and processing software was written for the older machines, but the graphics were crude and only limited image processing was possible without dedicated hardware. Although networking was possible to an LSI 11, it is integral to a workstation. Thus, the EDS system could be networked to a central computing facility or to other computers in the microscopy laboratory.

PCs of the time were nearly as fast as workstations, but constraints of their operating system limited the amount of data that could be handled as well as the number of processes that could be run. Their display was also limiting. Workstations typically have one million or more pixels in the display, facilitating the display of multiple images simultaneously. It is also possible to expand the X-ray display to examine many spectra at the same time or to shrink the display to carry out other operations while spectra are acquiring in the background. A very large number of colors are also available to display data in a way that makes differences obvious.

The 1990s

Detector

The field seems to have settled on atmospheric thin window detectors. Few new Be-window detectors are produced, though there are many still in the field. Again, few new windowless detectors are made, even though they still offer the ultimate light-element sensitivity.

Improvements in spectral resolution have resulted from advances in silicon crystal processing. But with resolutions at manganese now in the low 130 eV range, little more improvement is possible. The resolution of a detector is determined by noise in the system and by statistical considerations resulting from random fluctuations in the number of charge carriers produced by a given energy deposited in the crystal. This latter statistical term is the limiting factor determining resolution. If a system were perfectly noise free, its resolution would still be in the low 120s, depending slightly on one's choice of values for the Fano factor for Si(Li).

Germanium Detectors. Work continues on germanium detectors, because their inherent resolution is better than that of Si(Li). The charge collection problems that limited light element performance in the early 1980s have been solved, and some Ge detectors have been produced with atmospheric thin windows for light element detection. For a thorough and up-to-date review of germanium detectors in EDS applications, see Sareen (13).

Counterbalancing germanium's better spectral resolution and high efficiency for heavy elements are several factors. One of these is that the surface of germanium detectors is less robust than that of Si(Li). The oxides of germanium are not as stable as that of silicon, and the leakage current is often high. Moreover, germanium is sensitive to infrared radiation and must be protected by a cryogenically cooled filter in front of the crystal. This filter reduces the efficiency of soft X-ray detection from light elements (see Figure 2). As a consequence, germanium is not the dominant detector material. Si(Li) is still used routinely for EDS, while the most common use of germanium is on a TEM, particularly an intermediate-voltage one (200-400 kV), because of its high-energy efficiency.

Digital Pulse Processing. Probably the most significant breakthrough in EDS hardware in recent times has been the digital pulse processor. The reason for this is that it frees the pulse processing chain from fixed-shaping analog circuitry. If the pulse stream coming out of the detector (preamplifier) is digitized at high speed, it can be buffered so that it is possible to anticipate the pulses. Unlike analog circuitry that must apply filters to detect and measure each pulse starting when it arrives, a digital system can begin the filtering process on the data stream before the pulse as well as after it. So it is possible to vary the shaping time to fit the time available. At low count rates, shaping is long, and the best energy resolution is achieved. At high count rates, the hardware adapts as necessary. But even at high count rates, the arrival of photons is a

Poisson distribution, so some long intervals occur, and for these, the hardware uses the longer time.

This is shown schematically in Figure 6. The figure compares fixed and adaptive shaping. Five separate pulses are depicted at various energies and times of arrival in the X-ray detector. Step height corresponds to energy, with small rises representing soft X rays and large rises corresponding to more energetic X rays. The top trace is the actual pulse stream (digital or analog), and the middle and bottom traces represent the response of an analog amplifier or digital pulse processor, respectively. Note the equally spaced tick marks on the analog trace compared with the variable widths on the digital trace. The time between these marks represents the pulse processing time.

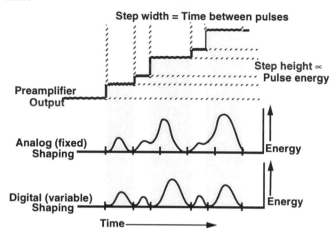

Fig. 6 Analog vs. digital shaping of the detector output

The analog system measures only one of the five pulses, the first. The digital system measures them all. Although the signal returns to zero after the first, third, and fifth pulses, the third and fifth are missed by the analog system, because the signal had not returned to baseline by the time they arrived. In the digital system not only does the shaping time adapt on the fly, but it is not even constrained to use the same time on either side of the event. The magnitude of the measuring filters can be changed to compensate for the time over which they are applied. This is called asymmetric adaptive shaping.

The practical consequence of digital processing is that, for a given resolution and input count rate, the digital system will collect an equal number of counts as analog in less than half the real time, thus increasing throughput. Or looking at it the other way, for the same real time, the digital spectrum will have about 2.5 times as many counts in it as in the analog spectrum. For a review of digital pulse processing as applied to metallurgy, see Friel and Mott (14), and a more thorough description of the technology can also be found (15).

In a digital system, the shaping time is continuously variable. This effect is shown in Figure 7. This set of spectra were taken with a modern germanium detector with digital pulse processing. The beam current on the microscope was adjusted to produce count rates of 2000, 5000, 10,000, and 30,000 cps at various resulting dead times. The spectra are of the same AgPd alloy shown in Figure 1 from 1976. Note that at up to 10,000 cps, the peaks of the two elements are resolved. And even at 30,000, peak deconvolution software was able to separate them as it did 20 years ago. Quantitative analysis of this Ag30Pd alloy was never in error more than 2 wt% at any count rate.

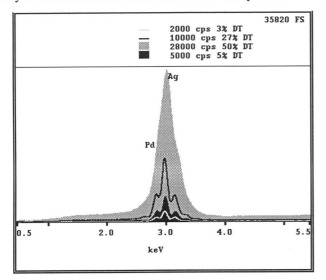

Fig. 7 Spectra of AgPd using digital processing

Analyzer

The host computers for EDS now include both workstations and PCs, and all with much less dedicated hardware than in earlier systems. Ease of use and automation have become a consideration, as EDS systems are sometimes entrusted to less experienced operators. As windowing systems converge on a similar look and feel, there is becoming less distinction among classes of computers in terms of their user interface.

Compositional Imaging. One way in which faster computers on EDS systems have been employed recently is compositional imaging. Even old analog maps might be considered compositional images, but the term really applies to digital images. These are images that can be processed, e.g., background subtracted; they can be compared as in image math; or they can be analyzed as in measurement of volume fraction based on chemistry. The term has also been used for digital maps that have had a matrix correction applied to convert intensity to weight percent. A series of papers on some of these aspects can be found in a publication of the Microscopy Society of America (16). The papers examine several signals that can be used for compositional imaging and discuss quantifying composition.

One of the most valuable capabilities of modern computers that relates to X-ray images is probably the ability to compare many images—from the Boolean comparisons discussed previously to quantitative image correlation. It is not difficult for a computer to calculate a correlation coefficient between all pairs of images on a pixel-by-pixel basis. For example, if one were looking for associations in a series of elemental maps and computed images, a matrix of correlation coefficients could be calculated to quantitatively make comparisons.

Position-tagged Spectrometry. Compositional images can be used to both gather information about spatial relationships and to display it. Analog dot maps were most useful for display only, whereas image math does both. The ability to collect an entire EDS spectrum at each pixel has been conceived of for many years, but it always taxed computer resources beyond their limit. Recently, Hunt and Williams described such a system for EELS and mentioned using it for EDS (17). They called it "spectrum imaging," and it consisted of collecting a full EDS spectrum at each pixel. In a different approach, instead of having the electron beam dwell for tens of milliseconds at each pixel as in spectrum imaging, one could scan the beam at full speed, recording the location from which each X-ray photon is detected. In other words, the photons measured by the EDS detector carry tags describing the position on the specimen from which they were generated.

The display can be an image, a spectrum of the scanned area, digital elemental maps or all of the above—each developing with time. Spatial resolutions of the electron image, displayed maps and stored maps can all be different, but the position tags carry the full resolution of the electron image.

In this position-tagged spectrometry (PTS), maps can be used not only to display spatial data but to gather it. Elements not selected originally can be displayed as maps after-the-fact or during acquisition. An area can be scanned just long enough to decide if it is suitable for a longer map. Even though the file generated by a high-resolution scan is quite large, the computer can scan the entire three-dimensional dataset (x, y, energy) for information not anticipated when the scan was begun. Now X-ray maps truly can be used to gather information about a specimen, not just display what is already known.

Conclusions

There have been many developments in EDS since the technique's inception. X-ray detectors have improved dramatically in spectral resolution, throughput, and the ability to detect and measure light elements. One now has a choice of detectors to meet more closely the analytical requirements of the user. Computer-based analyzers offer both the power of modern processors as well as the sophisticated software developed specifically for X-ray microanalysis and computer-aided imaging on an SEM. Particular progress has been made since digital X-ray maps became the standard. Computer processing of X-ray maps now reveals information about the specimen that would be difficult or impossible to retrieve otherwise. And most recently, the application of digital pulse processing to EDS illustrates that more innovations are not only possible, but likely.

Acknowledgement

The author wishes to acknowledge the technical editing skill of Marie Kuszewski and her valuable assistance in the production of this manuscript.

References

1. Fitzgerald, R., K. Keil, and K. F. J. Heinrich, Science, 159, 528-529 (1968).
2. Fischberg, A., Applied Sciences Associates, personal communication (1995).
3. Jaklevic, J. M. and F. S. Goulding, IEEE Trans. Nucl. Sci., 18, 187-191 (1977).
4. Russ, J. C. and A. O. Sandborg, "Energy Dispersive X-ray Spectroscopy, NBS Special Publication 604," National Bureau of Standards, 71-95 (1981).
5. Myklbust, R. L., C. E. Fiori, and K. F. J. Heinrich, in Proceedings of the Eighth Int. Conf. on X-ray Optics and Microanalysis, 96A-96D (1977).
6. Schamber, F. H., in Proceedings of the Eighth National Conference on Electron Probe Analysis, New Orleans, Paper 85 (1973).
7. McCarthy, J. J., J. J. Christenson, and J. J. Friel, American Laboratory, 9-15 (1977).
8. Barbi, N. C., and D. B. Lister, "Energy Dispersive X-ray Spectroscopy, NBS Special Publication 604," National Bureau of Standards, 35-44 (1981).
9. Cox, C. E., B. G. Lowe, and R. A. Sareen, IEEE Trans. Nucl. Sci, 35, 28-33 (1988).
10. Aden, G., and D. Isaacs, Research & Development, 8, 85-90. (1987).
11. Packwood, R. H., and J. D. Brown, X-ray Spectrometry, 10, 138-146, (1981).
12. E 1508, "Annual Book of Standards," 03.01 American Society for Testing and Materials, Philadelphia, PA
13. Sareen, R. A. "X-ray Spectrometry in Electron Beam Instruments," p 33, Plenum Press, New York (1995).
14. Friel, J. J., and R. B. Mott, Advanced Materials and Processes, 145, 35-38 (1994).
15. Mott, R. B. and J. J. Friel, "X-ray Spectrometry in Electron Beam Instruments," p 127, Plenum Press, New York (1995).
16. Lyman, C. E., "Microscopy, The Key Research Tool," p 1, Electron Microscopy Society of America (1992).
17. Hunt, J. A., and D. B. Williams, Ultramicroscopy, 38, 47-73, (1991).

Development of Backscattered Electron Kikuchi Patterns
for Phase Identification in the SEM

J. R. Michael and R.P. Goehner
Sandia National Laboratory, Albuquerque, NM

Abstract

This paper describes the use of backscattered electron Kikuchi patterns (BEKP) for phase identification in the scanning electron microscope (SEM). The origin of BEKP is described followed by a discussion of detectors capable of recording high quality patterns. In this study a new detector based on charge coupled device technology is described. Identification of unknown phases is demonstrated on prepared and as received sample surfaces. Identification through a combination of energy dispersive x-ray spectrometry (EDS) and BEKP of a Laves phase in a weld in an alloy of Fe-Co-Ni-Cr-Nb and the identification of $Pb_2Ru_2O_{6.5}$ crystals on PZT is demonstrated. Crystallographic phase analysis of micron sized phases in the SEM is a powerful new tool for materials characterization.

THE IDENTIFICATION OF UNKNOWN MICRON-SIZED PHASES in the SEM has been limited by the lack of a robust and simple way to obtain crystallographic information about the unknown while observing the microstructure of the specimen. A variety of techniques are available that can provide some information about the identity of unknown phases. For example, EDS is of some use but obviously cannot distinguish between phases with similar compositions but different crystal structures (an example of this is TiO_2 that has two tetragonal forms with different atomic arrangements and an orthorhombic form). Electron channeling patterns (ECP) in the SEM can provide appropriate information about the crystallography of the unknown, but the technique is limited by spatial resolution and sensitivity to plastic strain and is not applicable to rough surfaces. Other techniques can provide the required information but have significant disadvantages when compared to BEKP. Selected area electron diffraction in the transmission electron microscope (TEM) can provide crystallographic information from micron-sized regions of the specimen, but TEM requires electron transparent thin specimens to be produced which is time consuming and can be very difficult. BEKP in the SEM can provide crystallographic information from sub-micron sized areas with little or no difficult specimen preparation.

Backscattered electron Kikuchi patterns were first observed over 40 years ago before the development of the SEM (1). The patterns were recorded on photographic film inside an evacuated chamber. It was not until the 1970's that a suitable detector system was attached to an SEM and BEKPs observed (2). At this time the authors recognized that the patterns could be used to determine accurate crystallographic orientations of sub-micron sized crystals (3). More recently, the technique has been highly developed to provide accurate crystallographic orientation of grains in polycrystalline materials. This information is then used to determine the microtexture of the specimen (4,5).

BEKP's have been used to identify crystal symmetry elements and crystallographic point groups. This work showed that 27 of the 32 possible point groups could be identified using BEKP (6-10). The pattern quality required for these studies dictated that photographic film be used to record the images. The recording of BEKP on film requires that a suitable film transport mechanism be added to the SEM, although recently a 35 mm camera body has been used inside the vacuum of an SEM for this purpose. Although, photographic recording of BEKP is possible, it is not a real time technique suitable for on-line phase identification.

The development of a camera for BEKP recording that is based on CCD (charge coupled device) technology has enabled phase identification to be performed rapidly in the SEM using BEKP (11). This paper will briefly discuss the origin of the patterns and then will describe how BEKPs can be utilized to provide crystallographic phase identification of sub-micron areas.

Origin of Backscattered Electron Kikuchi Patterns

Backscattered electron Kikuchi patterns are obtained by illuminating a highly tilted specimen with a stationary electron beam. The beam electrons are elastically and inelastically scattered within the specimen with some of the electrons scattered out of the specimen. These backscattered electrons form the diffraction pattern. There are currently two mechanisms used to describe the pattern formation. In the first description, Kikuchi patterns are formed by the elastic scattering (diffraction) of previously inelastically scattered electrons. Some of the inelastically scattered electrons satisfy the Bragg condition for diffraction and are diffracted into cones of intensity which form the pattern when imaged with a suitable detector (1). An alternative description of the pattern formation requires one high angle scattering event that results in an electron exiting the specimen. The channeling of these electrons by the crystallographic planes of the specimen results in the formation of the cones of intensity in a matter analogous to the channeling of electrons in electron channeling patterns (12). Fortunately, we do not require a detailed understanding of the physics of electron scattering to use these patterns for phase identification in the SEM.

The specimen must be tilted to a high angle (typically 70°) with respect to the electron beam for two reasons as shown in Figure 1. Figure 1 is a plot of the energy distribution of backscattered electrons from Ni for a beam voltage of 30 kV for an electron beam normal to the specimen surface and for a beam that is 70° from the surface normal. The advantages of utilizing a tilted specimen can be shown by comparing the data for the normal incidence and the tilted beam. First, the number of backscattered electrons is greatly increased by tilting the sample. This results in a much greater signal that permits the patterns to be observed.

Secondly, when the specimen is tilted to high angles, the backscattered electrons do not penetrate as deeply into the specimen as for the untilted specimen. Thus, the backscattered electrons lose a much smaller amount of energy on average resulting in a sharp peak in the backscattered electron energy distribution that is very close to the initial beam energy. The electrons in this sharp peak contribute to the diffraction pattern. The electrons that are contained in the sharp peak have not traveled to great depths into the specimen, thus BEKP is a technique that samples a relatively thin surface layer. Lower accelerating voltages will result in a signal that is more highly localized to the surface of the specimen. The electrons that have experienced larger energy losses contribute to the overall background intensity in the patterns.

The diffracted electrons form cones of intensity. The cone axis is normal to the crystal planes that satisfied the Bragg condition and the cone semi-angle is 90° minus the Bragg angle. Since the Bragg angle is relatively small (about 3°) the cones are very shallow (12). The diffracted cones are then detected with an appropriate detector. When the cone of intensity intersects a planar detector surface, a conic section is produced. However, since the cone angle is large, the conic sections are nearly straight lines and appear curved only at low beam energies where the Bragg angle is larger.

Information Available From BEKP

Figure 2 shows a BEKP obtained from a metallographically polished Ni sample that was recorded using an accelerating voltage of 20 kV and a slow scan CCD camera. The pattern consists of a large number of parallel lines. Each line pair is the result of diffraction from the front and back of a particular plane in the crystal.

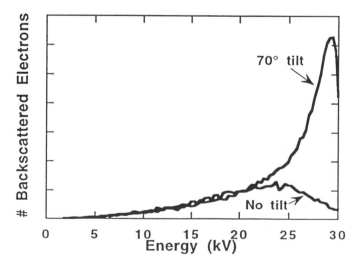

Fig. 1 Plot of the energy distribution for electrons backscattered from a tilted and a flat sample in the SEM.

Fig. 2 BEKP of metallographically prepared Ni recorded at 20 kV.

The angular width between each line pair is proportional to the interplanar spacing that gives rise to the line pair. The pattern also contains a number of prominent areas where the pairs of lines intersect. These intersections are zone axes and the angle between the various zone axes is indicative of the samples crystal structure.

BEKP's can be obtained for a range of accelerating voltages (and therefore electron wavelengths). Figure 3 a and b shows patterns obtained from Si at 40 and 10 kV. The positions of the zone axes remain fixed as the angular relationship between zone axes is determined by the crystal structure of the sample and not the beam voltage. The width of the line pairs vary with the accelerating voltage as described by Bragg's Law. As the wavelength of the electrons increases (lower accelerating voltages) the Bragg angle increases. Thus the line pairs will have a larger angular spacing between them. This is apparent when Figure 3a is compared to Figure 3b. The increased curvature of the line pairs at lower accelerating voltages is also apparent in Figure 3b along with the reduced definition of the lines at the lower operating voltage due to the larger energy loss suffered by the electrons.

It is fairly easy to calculate the orientation of the specimen from a single pattern (13,14). This is the basis of the technique of orientation imaging microscopy (OIM) where the orientation of a microstructure is mapped by obtaining patterns at each point in an array of points on the sample surface and then using orientation differences between the patterns to delineate microstructural features (14-17).

Collection of Backscattered Electron Kikuchi Patterns

BEKPs can be successfully collected using photographic film, video rate camera technology or slow-scan CCD technology. Although high quality patterns can be collected on photographic film, film suffers from the disadvantage that the collected image can be viewed only after the film is developed. BEKP's have been collected with video rate cameras. In this arrangement the diffracted electrons from the sample strike a scintillator that converts the electrons to light. The scintillator is then viewed with a video rate low light level camera to produce an image on a TV or computer screen. Patterns collected in this manner are suitable for the determination of sample orientation, but are not of sufficiently high quality for phase identification.

In order to overcome the disadvantages of photographic film and video rate recording of the patterns, a camera was developed based on CCD technology. This camera consists of a thin yttrium aluminum garnet (YAG) scintillator that is fiber optically coupled to a scientific-grade thermoelectrically-cooled slow-scan CCD. A CCD with 1024X1024 pixels was chosen for this application to achieve high accuracy in the measurement of the recorded BEKP. In order to achieve a large collection angle the CCD, 2.54 cm X 2.54 cm, was coupled to the scintillator through a 2.5:1 fiber optic

reducing bundle. This detector configuration permits high quality patterns to be recorded using exposures of 1 to 10 seconds (11). The actual exposure depends on the specimen (atomic number) and the electron optical conditions (accelerating voltage and beam current).

The CCD is a good replacement for photographic film recording of BEKP's. Typically, high quality CCDs have a higher dynamic range than photographic film, but the resolution of the CCD still cannot approach the resolution of photographic film (18,19). However, for images collected at a resolution of 1024X1024, it is difficult for the unaided eye to resolve the individual pixels. For phase identification using BEKP in the SEM, CCD detectors are superior to film due to the high dynamic range of the CCD and their adequate resolution and the immediate availability of the image for analysis (19).

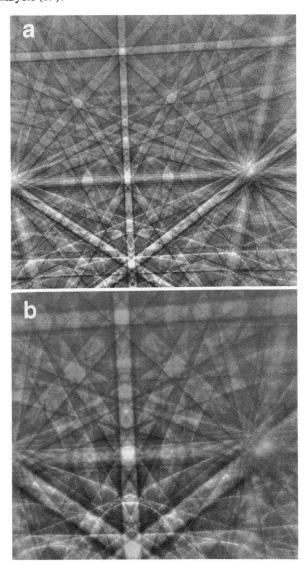

Fig. 3. BEKP of Si obtained in the SEM using CCD camera. a) 40 kV, b) 10 kV.

The patterns obtained are improved significantly by an image processing procedure called flat fielding (18). The raw patterns collected with the CCD consist of relatively weak but detailed crystallographic information superimposed on a slowly changing background intensity and image artifacts related to the fiber optic bundle and the scintillator. Once a raw image has been recorded a second image (the flat field image) is recorded that contains everything but the crystallographic information. This image may be obtained by scanning the beam over a number of crystals while collecting a pattern or by recording patterns from very fine grained materials. The image artifacts and the slowly varying background component can be removed by normalizing the raw image containing the crystallographic information with the flat field image. This procedure has the added advantage of increased contrast in the processed patterns. An example of the flat fielding procedure is shown in Figure 4 a and b.

The accurate determination of crystallographic parameters from BEKPs requires that the camera system be carefully calibrated (15). The two parameters that require calibration are the pattern center and the specimen to detector distance. The pattern center is defined as the projection of the beam impact point onto the scintillator. There are a number of ways to calibrate the pattern center. We have found that one of the most accurate and simple ways to accomplish this is to record patterns from the same crystal at two different scintillator to specimen distances (11). The crystallographic zones move along radial lines that project out from the pattern center. It is quite simple to plot the positions of the zones as they move to determine the pattern center. This technique has been used to determine the pattern center to within one or two pixels. The CCD camera has been calibrated using this procedure and has been found to be extremely stable with time. Once the pattern center is accurately known the specimen to detector distance can be easily determined from a known specimen. Table 1 shows the measured angles between zones in Figure 3a and compares the measured values with values calculated from the crystal structure.

Table I. Comparison of measured interzonal angles with actual angles for Si

Angle between	Measured (°)	Actual (°)
[001] - [111]	54.81	54.7
[001] - [011]	44.93	45.0
[011] - [112]	29.94	30.0

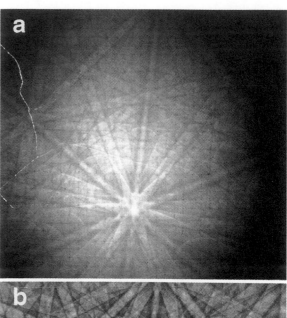
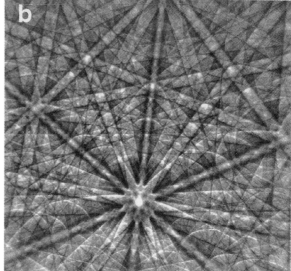

Fig. 4 Flat fielding procedure can greatly increase contrast and quality of BEKP images. a) raw image before flat fielding b) image after flat fielding.

Procedure for Phase Identification using BEKP

Phase identification with BEKP in the SEM has previously been accomplished by recording patterns on film. On-line phase identification has not been possible due to the necessity to expose and develop the photographic film. The CCD detector system eliminates this problem, as the patterns are immediately available for viewing and analysis and are of similar quality to those obtained on photographic film. Phase identification is achieved in the following manner.

First an area of interest is located on the tilted specimen using the appropriate imaging signal (backscattered or secondary electrons). A BEKP is then collected from the feature of interest using a stationary beam in the same way that x-ray information is obtained. A flat field image is then obtained and used to flat field the raw image data as has been described previously. Compositional data is also obtained from the feature of interest as this information is used to limit the number of possible candidate materials that could produce the BEKP.

The next step in phase identification using BEKP is to determine the angles between the various zone axes present in the image. Candidate crystal structures, consistent with the compositional information obtained from the unknown, are selected from the literature. The measured interzonal angles are then compared with those calculated from the candidate crystal structure. Quite often these angles and the compositional information are sufficient to unequivocally identify the unknown crystalline phase. Once a possible match is found, the pattern is simulated in the correct orientation. The simulation includes the effect of structure factors and calculates the relative intensity for each reflection. The simulated pattern is compared to the experimentally obtained pattern and if the agreement between the experimental and the simulated pattern is good, then an identification is assumed to have been accomplished.

It should be noted that in a vast majority of cases there is no need to make use of the interplanar spacings that can be measured from the BEKPs. Cubic materials present the most difficulty in using the approach described above because the interzonal angles do not vary with the lattice spacing of the crystal. As the symmetry of the crystal decreases the angles between the zone axes in the pattern becomes more diagnostic. Fortunately, it is possible to measure the interplanar spacings from a BEKP. This is done by measuring the angle between the pair of lines that produce the Kikuchi band and converting this angle to a interplanar spacing (14). This data can then be used as a further condition for identifying the unknown phase of interest and is particularly useful for cubic crystal structures.

Examples of Phase Identification

Identification of Laves Phase in Welds

During the simulated welding of an Fe-Co-Ni-Cr-Nb alloy a second phase was observed in metallographically prepared sections. Figure 5 is a backscattered electron image of the simulated weld that shows the precipitates that image brighter than the matrix phase. EDS x-ray analysis in the SEM demonstrated that the second phase was rich in Fe and Nb.

BEKPs of the bright imaging phase were obtained from a metallographically polished, but not etched sample. The SEM was operated at 30 kv with a beam current of 1-2 nA. Figure 6 shows an example of the BEKP from the bright precipitates in the simulated weld. A survey of the literature showed that there were a number of phases that contain Fe

and Nb. Possible Nb-Fe phases were NbFe, Fe_2Nb and Fe_7Nb_6 (20). Interzonal angles observed in the experimental patterns were measured and compared with the calculated values for the above three compounds. Good agreement was found between the interzonal angles for Fe_2Nb and the experimentally measured interzonal angles. Patterns for Fe_2Nb were simulated, including the effects of structure factors, and the simulation was compared with the experimental patterns. Figure 7 is a BEKP simulation for Fe_2Nb calculated using the same microscope parameters as those used for collecting the BEKP from the unknown bright phase. As can be seen from comparison of Figure 6 and Figure 7 the agreement is excellent. Thus, the bright phase was identified as Fe_2Nb, a Laves phase that is hexagonal (a=.4837 nm and c=.7884 nm).

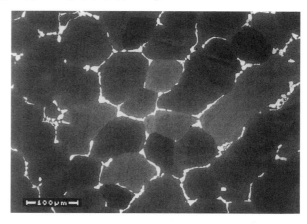

Fig. 5. Backscattered electron image of weld simulation in Fe-Co-Ni-Cr-Nb alloy.

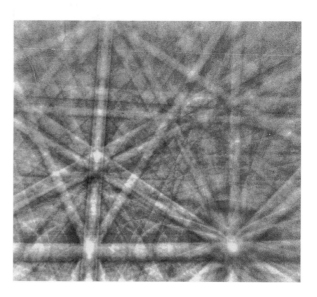

Fig. 6. BEKP obtained from the bright phase shown in Figure 5.

Fig. 7 Simulated BEKP of Fe_2Nb that demonstrates excellent agreement with Figure 6.

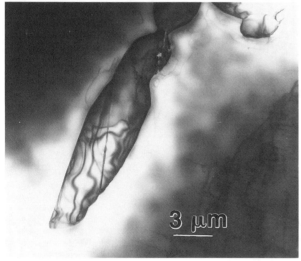

Fig. 8 Bright field TEM micrograph of the bright phase shown in Figure 5.

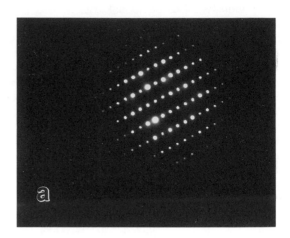

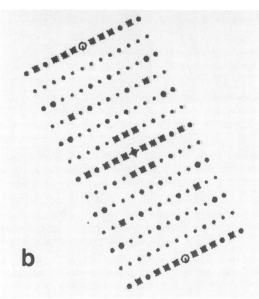

Fig. 9 Experimental and simulated SAED patterns from the Laves phase (Fe_2Nb) shown in Figure 8. a) experimental, b) simulated [2 -1 -1 0] zone.

In order to verify and demonstrate the correct identification of the bright phase as Fe_2Nb, selected area electron diffraction (SAED) was performed in the transmission electron microscope. SAED in the TEM requires that thin foil samples be produced from the bulk weld material. This can be a tedious and time intensive procedure and is particularly so when samples with large second phase precipitates are to be produced. In the case of the Fe-Co-Ni-Cr-Nb alloy electropolishing was used to produce electron transparent specimens for SAED analysis. This demonstrates one of the main advantages of BEKP for phase identification which is the lack of time consuming specimen preparation. Figure 8 is a TEM bright field micrograph of the bright phase shown in Figure 5. In order to obtain good indexable SAED patterns in the TEM it is necessary to tilt the specimen

carefully to obtain a zone axis pattern. The limited angular view of SAED in TEM requires that the specimen be tilted to a variety of zone axes so that the phase may be identified. Figure 9a is a SAED pattern of the phase shown in Figure 8. Figure 9b shows a simulation of the SAED shown in Figure 9a. In order for a complete analysis a number of patterns were collected and indexed. The SAED analysis in the TEM demonstrated that the second phase in the simulated weld was indeed Fe_2Nb confirming the conclusions of the BEKP analysis. The limited angular view provided by SAED in the TEM is a distinct disadvantage of the technique that requires multiple patterns to identifiy a phase. The large angular view provided by BEKP permits the collection of data from a large portion of the stereographic triangle in only one

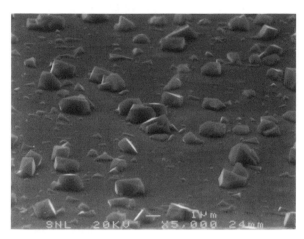

Fig. 10. Secondary electron image of the crystals on the PZT thin film.

exposure thus obviating the need to tilt to obtain different patterns.

Identification of Crystals on PZT Thin Films

The previous example demonstrate the applicability of phase identification using BEKP to metallographically prepared specimens. This example demonstrates that the technique can also be used on rough surfaces. In this example BEKP was used to identify crystals that had formed during the thin film processing of lead zirconium titanate (PZT) thin films. The processing involved the deposition of thin films of RuO_2 onto Si wafers followed by the deposition of the PZT thin film. After deposition of the PZT, crystals on the surfaces of the processed wafers were noted. Figure 10 is a SEM photograph of the crystals on the PZT films. The crystals were on the order of 1-2 μm in size and were distributed across the wafer surface. In this case it would be quite difficult to prepare suitable samples for TEM analysis so identification was attempted using BEKP.

Figure 11 is a BEKP obtained from a crystal shown in Figure 10. The pattern is very clear and well defined, even though it was obtained from a non-flat specimen that had been coated with carbon for conductivity. Qualitative EDS analysis of the crystals showed them to contain Pb, Ru, and possibly O. The interzonal angles were measured and are shown in Table 2 and indicate that the crystals structure was cubic, i.e. that the three lattice parameters are equal and at 90° to each other. Table 2 also shows a comparison of the experimental values with those calculated for a cubic crystal structure. Examination of the literature indicated that there were no oxides of Pb or Ru that were cubic, however there is a cubic pyrochlore phase $Pb_2Ru_2O_{6.5}$ that had been reported. It is important to point out that the interzonal angles

Fig. 11. BEKP obtained from a crystal shown in Figure 10. The pattern was obtained at 40 kv.

Table II. Comparison of interzonal angles from BEKP of crystals on PZT with angles calculated for cubic compounds

Angle Between	Measured (°)±0.3°	Computed (°)
[110] - [310]	26.6	26.6
[110] - [211]	29.9	30.0
[110] - [121]	30.4	30.0
[110] - [111]	35.2	35.26
[111] - [121]	19.7	19.5
[100] - [110]	44.5	45.0

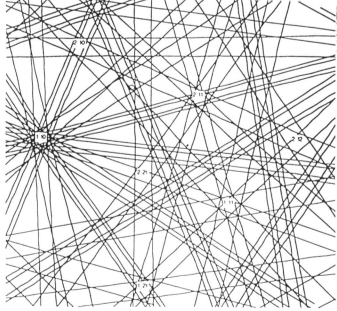

Fig. 12 Simulated pattern of $Pb_2Ru_2O_{6.5}$ that shows excellent agreement with the experimental pattern shown in Figure 11.

indicate only that the crystal structure is cubic, all cubic materials will have the same interzonal angles. Unequivocal identification required that the d-spacing be measured and compared with those for $Pb_2Ru_2O_{6.5}$. Table 3 shows the d-spacings measured from the pattern shown in Figure 11. The measured d-spacings are in good agreement with the reported values for $Pb_2Ru_2O_{6.5}$. Although, in this case the agreement is quite good, further development is needed to routinely extract accurate d-spacing measurement from these patterns.

Table III. Comparison of measured and computed d-spacings for $Pb_2Ru_2O_{6.5}$

Plane (hkl)	Measured d (nm)	Computed d(nm) a=b=c=1.025 nm
(222)	0.30±0.02	0.2959
(400)	0.25±0.01	0.2563
(440)	0.18±0.01	0.1812
(220)	0.15±0.01	0.1546

The final step in the identification of the crystals as $Pb_2Ru_2O_{6.5}$ is to simulate the BEKP. Figure 12 shows the simulation of the pattern. Comparison of this simulation with the experimental pattern shown in Figure 11 demonstrates excellent agreement and thus the crystals that grew on the PZT film were identified as $Pb_2Ru_2O_{6.5}$.

Summary

The combination of BEKP and EDS now permits the unequivocal identification of micron sized areas in the SEM. The ability to obtain high quality BEKPs without the use of photographic film enables the patterns to be analyzed rapidly. No other technique can as quickly and accurately provide the identification of the crystallography of micron-size areas of a specimen. In this paper we have described the origin of and the information contained in these patterns. This paper has demonstrated the crystallographic identification using BEKP of unknown phases in metallographically prepared samples and of unprepared surfaces. Although unequivocal identification is not yet possible in the general case, the combination of diffraction information from BEKP and compositional information from EDS is a powerful tool for materials characterization.

Acknowledgments

The authors would like to thank Dr. B. A. Tuttle for supplying the specimens of PZT and Dr. C. V. Robino for the specimens of the simulated weld. The authors would also like to thank Dr. T. J. Headley for his careful review of this paper. This work was supported by the U.S. Department of Energy under contract #DE-AC04-94AL85000.

References

1 Alam, M. N., M. Blackman and D. W. Pashley, Proc. Roy. Soc., 221, 224 (1954).
2 Venables, J. A. and C. J. Harland, Phil. Mag., 27,1193 (1973).
3 Venables, J. A. and R. Bin-Jaya, Phil. Mag., 35,1317 (1977).
4 Dingley, D. J. SEM, IV, 273 (1981).
5 Dingley, D. J. SEM, II, 569 (1984.
6 Dingley, D. J. and K. Baba-Kishi, SEM, II,383 (1986)
7 Baba-Kishi, K. Z. and D. J. Dingley, Scanning, 11, 305 (1989).
8 Baba-Kishi, K. Z. and D. J. Dingley, J. Appl. Cryst., 22, 89 (1989).
9 Baba-Kishi, K. Z. Ultramicroscopy, 36, 355 (1991).
10 Baba-Kishi, K. Z. J. Appl. Cryst., 24, 38 (1991).
11 Michael, J. R. and R. P. Goehner, MSA Bulletin, 23, 168 (1993).
12 L. Reimer, "Scanning Electron Microscopy," p 313,Springer-Verlag, New York (1985).
13 Harland, C. J. , P. Akhter, and J. A. Venables, J. Phys. E: Sci. Instrum., 14, 175 (1981).
14 Wright, S. I. and B. L. Adams, Metall. Trans. A., 23, 759 (1992).
15 Dingley, D. J. and V. J. Randle, Mater. Sci.,27, 4545 (1992).
16 Adams, B. L., S. I. Wright, and K. Kunze, Metall. Trans. A.,24A, 819 (1993).
17 Wright, S. I., B. L. Adams, and K. Kunze, Mat. Sci and Eng., A160, 229 (1993).
18 Kujawa, S. and D. Krahl, Ultramicroscopy, 46, 395 (1992).
19 Janesick, J. R., T. Elliot, S. Collins, M. M. Blouke and J. Freeman, Optical Engineering, 26, 692 (1987).
20 P. Villars, and L. D. Calvert, "Pearson's Handbook of Crystallographic Data," p 3303, ASM International, Metals Park, OH (1991).

Surface Analysis, Depth Profiling and Other Conventional Applications for the Electron Probe Microanalyzer

R. H. Packwood
Natural Resources Canada, Ottawa, Ontario, Canada

Guy Remond
Bureau de Recherches Géologiques et Minières, Orléans, France

Introduction

Castaing's original aim for the newly invented microprobe was to be able to analyze a sample volume of the order of 1×10^{-9} mm^3 with an accuracy of +/- 1%. It is fair to say that this ambition has been both met and exceeded.

The purpose of this paper is to emphasize that there is much more to be done with the instrument than regular analyses; things that have been forgotten or overlooked, which, taken together with more recent developments, show the microprobe to be a leading edge investigative tool for materials characterization.

- Analyzing layered samples and surface deposits with thicknesses from 0.1 monolayer to a few microns.
- Accurately quantifying non-uniform concentration distributions such as Ion-implanted semiconductors.
- Making depth profiles with resolution of 25nm or better.
- Measuring trace element concentrations into the 10ppm range.
- Light element analyses.

These are now all straightforward, indeed conventional applications for the microprobe and for some applications electronprobe microanalysis or EPMA may well be the method of choice.

How can this be? After all it is common knowledge that the microprobe cannot do those things! The answer of course is that the sources of information are wrong, and have been so, for a long time. We will discuss some of these claims in the next few pages. The literature cited will cover many more examples.

Advances in EPMA

The $\phi(\rho z)$ theory describing the generation of x-rays as a function of ρz, mass-depth into the specimen, is where the most dramatic improvements have been made. For example the simple MSG$\phi(\rho z)$ model can be used to investigate a wide range of sample configurations and location inside the specimen. Recent work (1) has vastly extended the basic notion so as to now permit accurate data prediction and therefore data correction for most of the Periodic Table, with the accelerating potential going from 1 to 100kev and specimen tilt angles up to 85°.

Similarly the PAP(2) correction procedure gives reliable results over the same range of elements and for all operating conditions normally encountered in the microprobe.

For specimens that are truly difficult and far removed from any resemblance to a semi-infinite, uniform solid, the Monte Carlo method still holds sway. The difference is that now, even on desk-top computers, the codes run very quickly so that thousands of trajectories per minute can be modeled. This in stark contrast with overnight runs on mainframe computers of yesteryear.

Correction procedures

As noted above, a large portion of in the advance in microprobe analysis has occurred in the area of data correction. The ZAF method is still available but the old limitations are unchanged. It cannot handle high x-ray absorptions or light elements except by comparisons between specimen and like standard. This is because the absorption correction in ZAF is often based upon Philibert's approximate formula and so is not appropriate for those circumstances. (To rectify the situation, at least in part, by adding a few lines of computer code to use Philibert's full formulation does not seem to have occurred to the software suppliers.)

To escape from these problems there are three main routes:-

i) PAP by Pouchou and Pichior. This is a large algorithm for correcting microprobe data, it models the primary x-ray generation with mass-depth by a pair of parabolic arcs spliced together so as to approximate the true x-ray distribution. The parameterization of the parabolas is based upon a large number of actual measurements and Monte Carlo simulations. This method will give accurate analyses in a wide range of experimental conditions. For the regular microprobe user this method produces very reliable data. Additional applications include tilted samples and layered specimens. On the other hand it is not particularly helpful when it comes to answering the sort of questions that are asked such as "What happens where?" inside the sample.

A variation on this theme has also been put forward by the same authors called XPP (3). It is particularly intended for use in the SEM. In place of parabolas two exponentials are used to model the x-ray generation volume. All the parameters in the $\phi(\rho z)$ formulation are expressed in terms of the specimen tilt angle.

ii)   The Monte Carlo method. This uses a computer program to calculate the individual electron trajectories through, and physical interactions inside, the mathematically modeled sample structure of choice; calculating x-ray production and exit path absorption as it goes. The great strength of this method is that, in principle, it can predict the observed signals from very complex structures. The trade-off is that the simulation of a significant number of electron tracks will take a few minutes, which is probably too long for use in a routine correction procedure.

To demonstrate the great versatility of the modern Monte Carlo method a small example is shown in Figure1 This commercial program took one minute to compute 1000 trajectories and the x-ray distribution. As yet the observed to generated x-ray intensity ratios are not reported but this would appear to be an oversight that can easily be remedied.

The topic is extensive and we cannot do justice to it in a review article format. The book by Joy (4) is strongly reccommended and the EPMA literature is a good source for up to date developments (5).

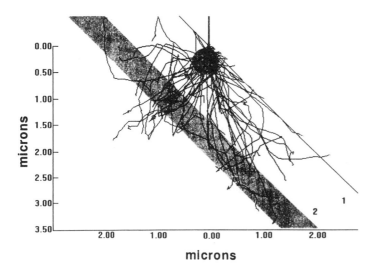

Fig. 1. Monte Carlo simulation of the interaction taking place in a multi-layer specimen. A sample comprised of an Fe particle 500nm diameter, three 500nm layers of Al, SiO2 and Si on a GaAs substrate tilted at 45° at 20kV is simulated

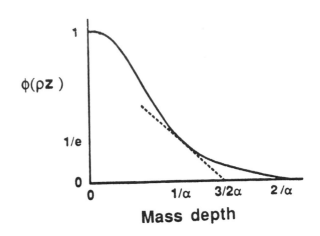

Fig. 2. Schematic of $\phi(\rho z)$ curve showing the meanings of extrapolated and ultimate ranges.

iii)   The modified surface-centered Gaussian or MSG$\phi(\rho z)$ This model can be looked upon as the random walk theory of x-ray generation in the microprobe. It is compact, with, in its simplest form, comparatively few equations but with results that are accurate and easy to interpret. For example the $3/2\alpha$ range is an extremely useful parameter, estimating as it does the mass-depth by which 85-90% of all x-rays are generated(6). See Figure 2.

The theory also makes predictions, the ultimate range, $2/\alpha$, for the electron flux is independent of the angle of incidence of the electron beam. Because it starts with electrons the theory can make predictions about electrons, for example, that the maximum depth from which a back scattered electron - BSE, can escape is one half the ultimate range and given by $2/2\alpha$. The 2α to be used depends on E_O, the energy of the incident electron beam, and the BSE detector's low energy cut-off. Single scattered BSE's are not included in this estimation. From layered specimens the BSE total is found to add as Error functions i.e. areas under sections of the appropriate Gaussian curves. This rule can be used in the STEM to measure ρt, the local foil thickness on the basis that the BSE flux varies as $Erf(2\alpha\rho t)$ as shown schematically in Figure 3.

To cover the entire field of microprobe applications would take several books. So instead we will describe in detail a few of the most useful methods and ideas with the intention of showing how simple and yet powerful this instrument has become. But first we take a short look at the theory of the microprobe.

Theory

EPMA is based on the assumption that the electron generated x-ray intensities are proportional to their elemental concentrations. Therefore we need to know how many x-rays are generated as a function of mass-depth in order to calculate the corrections for self-absorption and fluorescence. In that fluorescence is a long range phenomenon, the details of which are not

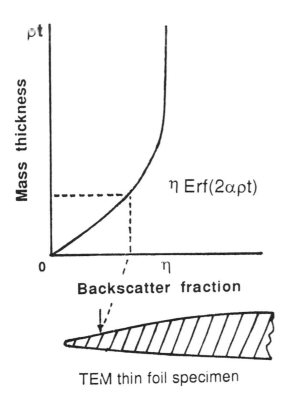

Fig. 3. Backscattering rate from thin film used to estimate local film thickness. Erf(2αρt) normalised to infinite thickness on massive region of specimen.

much altered by what follows, it is reasonable to say that standard Read (7) method will suffice. We note that fluorescence occurs when an x-ray of interest is excited by another, higher energy x-ray, generated by the electron beam. On the other hand, the absorption-atomic number or $\phi(\rho z)$ part of the calculation has been completely rethought and is worth reviewing at this time.

$\phi(\rho z)$ is the ratio of two x-ray intensities, that generated inside an ideally thin layer of the element of interest in or on the specimen, compared with that generated in the same layer held isolated in space. Curves of $\phi(\rho z)$ as a function of ρz can be measured or predicted by Monte Carlo calculations and there is a large data base available against which theoretical predictions can be tested. Once known these can be used to correct the observed intensities to find the underlying generated signal.
A brief outline of the MSG$\phi(\rho z)$ theory is given in the Appendix.

Layered Structures

The most elementary layered structure comprises a thin surface deposit of element A on a massive substrate of element B. This configuration occurs quite often in practice and so was the subject of some of the earliest microprobe work (8). Cockett and Davis gave estimated detection limits of the order of 1-2 monolayers on a surface. A simple order of magnitude calculation will illustrate the point. A typical x-ray generation volume in a medium atomic number matrix at 20keV electron beam

energy will be about 1micron across and 0.5micron deep. (It will not be pear shaped, that only holds for really low atomic number matrices, but more like a Vidalia onion!) This means around 2500 atomic layers are excited for a total of 3×10^{10} atoms. A single layer will have around 4×10^{7} atoms and so will generate a signal of 1/1000 that of the standard, i.e. 0.1%. This is not a challenge.

By following this same line of reasoning a step further, a monolayer of A on a grain boundary passing through the beam center would give 0.05% and a monolayer on two intersecting grain boundaries 0.075%. Of course to be useful the solid solubility of A in B should best be lower than these values.

A more accurate calculation yields the ratio of signal from surface layer, I(layer), to standard, I(std) as:-

$$I(layer)/I(std) = \phi_0 C_A.\Delta(\rho z) / \int_0^\infty \phi(\rho z).\exp(-\chi\rho z).d\rho z \qquad (1)$$

where the items on the right-hand side of the equation are the surface ionization, weight fraction of A, the mass thickness of the layer and the signal from the standard expressed as an integral. An additional factor allowing for the absorption in thicker layers can also be included.

As an example of the general principle Figure 4 shows a line trace for Ti-Kα over a cross-section through a graphite fiber coated with about 10nm of titanium diboride in an aluminium matrix. This is equivalent to approximately 25 atomic layers of titanium. The Ti-Kα signal is equivalent to 3wt% Ti roughly as predicted.

It must be emphasized that the microprobe does not have the spatial resolution to prove that the coating is 10nm thick but by suitable specimen preparation and using knowledge about the material at hand the instrument can be used to quantify such structures. Certainly an exposed grain boundary on a surface can be

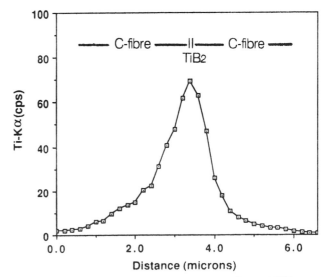

Fig. 4. Ti-Kα line trace over cross-section throughTiB2 coated graphite fibers. 10nm of coating yields signal equivalent to 3wt% Ti

examined before and after cleaning in some controlled fashion and thus demonstrate the presence or otherwise of suspected contaminants.

For more details on this topic see Packwood and Remond (9).

Depth Profiling

There are several ways to get information about the depth distribution of alloying or implanted elements. The first description of physical depth profiling with the microprobe was reported in 1984 (10). A SAM instrument was used to remove known amounts of surface material in between microprobe measurements on the area being cleaned. The key notion was separating the material removal aspect from the analytical step. Specimen transfer back and forth between instruments is not a problem; in the microprobe the layer of adsorbed water and hydrocarbons, that are a nuisance in true surface specific techniques, can almost be ignored. Neither the incident electron beam nor the escaping x-rays are appreciably affected.

Shoestring Sectioning is an inexpensive variation on the specimen dimpler machines used to prepare foils for TEM examination. An abrasive coated thread, constrained by guides, and under a tension of a few 100gm wt., is drawn back and forth across the area of interest on the specimen. A suitable choice of thread, grinding/polishing compound and tension yields grooves around 1mm wide and perhaps 0.5micron deep in a minute or so. If necessary the cross-sectional shape of the groove can be measured by surface profilometer.

The exposed subsurface layers in the walls of the groove can now be examined under the microprobe using a method that has been dubbed **SAXE** or Surface Analysis by X-ray Emission (9). The electron beam will lose energy as it travels into the body of the specimen and, if it starts out only just above E_C, the critical excitation potential, the $3/2\alpha$ range may be just a few 10's of nm. However to generate a reasonable x-ray intensity the overvoltage ratio U_O, defined as E_O/E_C, needs to be as large as possible, certainly well away from 1, because x-ray generation, as given by the ionization cross-section, is roughly proportional to $(1/U_O)\ln(U_O)$. In general this means that the soft x-ray part of the spectrum is used, and even then the count rate may be quite low. On the other hand the depth of excitation can be intriguingly shallow.

An application of this idea is shown in Figure 5, where a Ge-rich layer in an Si matrix has been exposed by shoestring sectioning. A depth resolution of around 25nm is displayed.

Using low U_O values has a second advantage. Not only does the x-ray volume stay close to the surface but the lateral spread of x-ray generation beyond the electron beam diameter is roughly twice the depth. An example of this effect is shown in Figure 6. Titanium-rich banding in a chlorite mineral is line scanned at 7kV. The observed Ti-Kα resolution is 0.5 micron, with about 0.25 micron due to the electron beam. Of course the bands

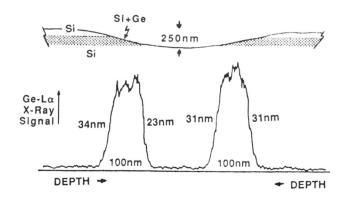

Fig. 5. SAXE. Top : section showing groove down into 100nm layer of Ge-rich Si in Si matrix. Bottom : Ge-Lα signal at 1.54keV vs Position on groove. Observed depth resolutions in nanometers noted beside peaks.

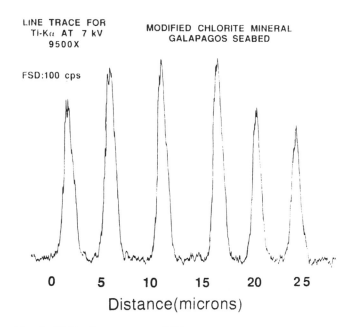

Fig. 6. Ti-Kα line trace at 7kV over cross-section through a specimen of banded chlorite from the Galapagos. X-ray spatial resolution (15-85%) of 0.5 micron is demonstrated.

themselves are not discrete and appear to have diffuse edges when seen at high magnification.

Taper sectioning For some specimens an adequate taper section can be achieved by conventional means, as with the galvanneal coating in Figure 7. A 6° slope gives a 10:1 'magnification' of the layered structure so that the electron beam can hit specific regions and, together with low U_O values, effectively localize the volume excited (11).

SIMS craters A second source of low angle taper sections are SIMS craters. In one instance a Au implanted silicon wafer was examined in the microprobe after SIMS work had been done (12). In Figure 8 the measured line trace from the top surface

74

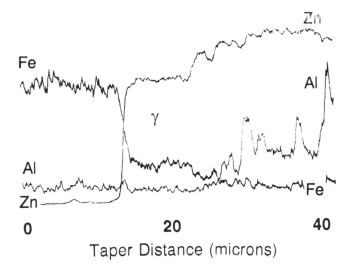

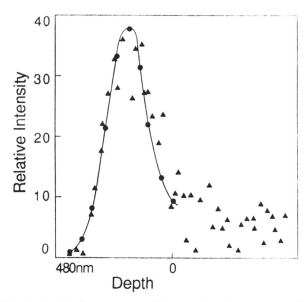

Fig. 7 Line traces for Fe-Kα and Zn-Lα at 10keV on a 6° taper - section through Fe-Zn interface on Galvannealed steel. Low penetration permits analysis of the γ layer which is around 0.5 micron thick.

Fig. 8 Au-Mα line trace at 5keV on Au ion implated Si, over SIMS crater edge down through implanted zone. Filled triangles are Au readings and the solid line the response predicted from theory and known ion implant profile.

down the crater wall compares well with the predictions of the MSGφ(ρz) theory. When large areas are available for investigation a series of craters of increasing depth can be made and then analyzed in the microprobe. Alternatively a sequence of crater-analyze steps would serve if the material available for investigation was limited.

One drawback with all large area depth profiling methods is that their final resolution is dependent upon the interface topography. A rough surface means that a low angle taper will be going up and down the

concentration levels. Only true high resolution techniques will have much chance of seeing the real interface behavior in this case.

From these ideas it is easy to turn to classical and non-classical taper sectioning techniques as a means to explore concentration changes that are relatively extensive at right angles to the depth dimension.

U_O depth profiling

Considerable effort has gone into the analysis of discrete layered structures by varying the accelerating potential and using mathematical models to retrieve the underlying concentrations. Programs such as Strata 2.01, PAP based (13), GMRTF by Waldo (14) or TFA by Bastien (15), both MSGφ(ρz) based, are available and simple to use. The idea behind these programs is shown in Figure 9. By starting at a relatively low E_O and working up by stages a series of extra 'shells' of signal are added, giving data that can be compared with prediction for progressively deeper layer thickness. Rickerby and Thiot (16) using Strata to analyse a Ti-BN multilayer structure report thickness estimates of 9.7nm and 15.2nm respectively. With maximum variations of around 10%, such values are equal to, or better than, the size of error found in AES sputter profiling.

If the approximate layer thicknesses are known then, rather than simply increasing the beam voltage by convenient steps, it could well be worthwhile estimating the various 3/2α ranges so as to concentrate the analyses at depths of most interest and thus get data that is really relevant to the sample being investigated.

Diffuse concentration variations

Many specimens have concentration distributions that are not discrete, for example a diffusion zone near a surface or ion-implanted semiconductors. The MSGφ(ρz)

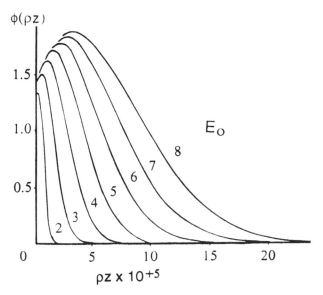

Fig. 9 φ(ρz) curves for As-Lα in Si as a function of accelerating voltage, Eo, illustrating principle of Uo depth profiling.

method can handle this situation quite well as seen above. The reason for this is simple, the MSG$\phi(\rho z)$ theory is based upon Gaussian and exponential functions and therefore is tailor-made for predicting x-ray intensities for concentration variations of those kinds. A simple substitution of coefficients in the standard formula is all that is required.

In fact the MSG$\phi(\rho z)$ theory can deal with a large variety of non-uniform concentrations in specimens. Figure10 is a schematic of the more commonly occurring distributions.

Trace element analysis

It is widely held that the microprobe cannot make trace analyses (17). However this is difficult to understand because it is an everyday event in our laboratories (18). As with many other techniques EPMA does not have a single minimum detection limit, or MDL, but a whole range that depends upon the particular combination of trace element and matrix being investigated. The light elements B, C, N and O all suffer in various ways either from high absorption factors in the specimen, or in the spectrometer counter windows, or from contamination that builds up around the electron beam impact site or oxidation from decontamination efforts. So for these elements MDL's will be quite large ranging from100 to1000ppm. For elements that only have L or M lines in the normal spectrometer wavelength interval of conventional microprobes, there is the problem of low count rates to contend with. However, these questions aside, there is still a large number of element pairs that can be measured at or near trace level concentrations i.e. around 10ppm.

The perception that the microprobe is not that sensitive may arise with naive users who will tend to stay with the short count times needed for conventional specimens:- typically 10-30sec at 20kV and 30nA beam current and software that does not printout the 4th decimal place. But even this explanation is flawed. Count rates of 10^4 -10^5 cps and peak to background ratios of 500:1 are commonplace giving 10^5 -10^6 counts per 10sec and a background count of 5000 in the same time. This latter is important because x-rays obey Poissonian statistics and so 50% of all background readings will fall within a range of plus and minus the square root of the background count +/-$\sqrt{}$(bkgd), here roughly equivalent to +/- 70ppm of a trace element. This on a single reading! In 100sec the error will be down to 20ppm. Of course more is generally required than a 50% likelihood; as a rule three times the background variation is used as a measure that approaches certainty. It should be evident that MDL's around10-20 ppm are possible in times of the order of 10 to 20 minutes.

To show that the method can handle real-life situations we show in Figure 11 a line scan for As and P across an input gate structure in a memory chip. The maximum concentrations are around 0.1wt%, the MDL suggested elsewhere!

Counting strategy. The basic principle here can be stated in one word - counts. The more counts the better, always assuming that the counter deadtimes do not accumulate too much and that the peak-to-background ratios are good. The only other concern is determining the background. Conventionally the background under an x-ray peak is estimated by linear interpolation from readings taken either side. However in general the real variation is not linear but decreases asymptotically with increasing wavelength. In this case the linear interpolation will overestimate the true background and give rise to negative peak intensities for the case of a specimen actually free of the element sought. The software handling the data correction will, unhelpfully, report these as zero concentrations. There are two difficulties here, the zeros can be avoided by taking a single offset reading on the long wavelength side and setting the 'slope' at 1.0. In most cases this will underestimate the background and yield false positive peaks, but they will be printed out for

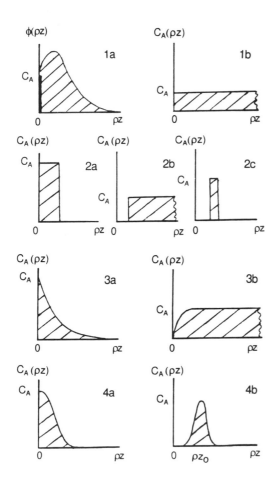

Fig. 10 Selection of concentration profiles that have formal solutions in the MSG$\phi(\rho z)$ theory. 1a - thin surface deposit with $\phi(\rho z)$ curve shown for reference, 1b - uniform, 2a - surface layer, 2b - buried substrate, 2c - buried layer, 3a - exponential, 3b - exponential depleted zone, 4a - Gaussian, 4b - implanted Gaussian.

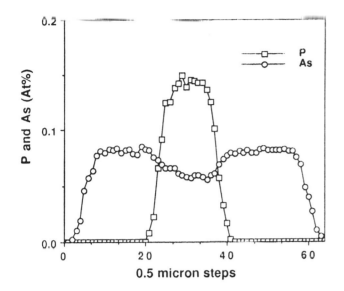

Fig. 11 Line traces for P-Kα and As-Lα at 5keV across input control structure of a memory chip. Measured concentrations are all less than supposed MDLs. At% used for clarity in center

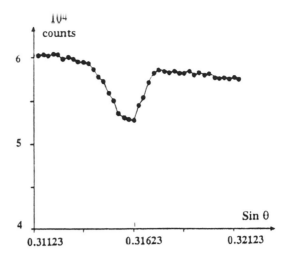

Fig. 12 Irregularities in X-ray continuum close to Au-Lα line. Recorded with an LiF crystal at 20keV.

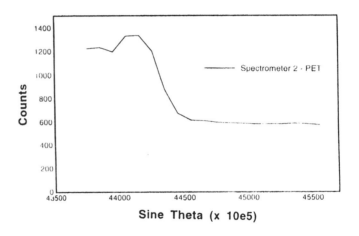

Fig. 13 Argon absorption edge in P-10 proportional counter gas adjacent to U-Mα line

inspection. For the second problem, the reading equivalent to zero can be measured on a blank specimen or calculated from readings at a second offset added to the first and the required background value calculated by ratio, assuming that the rate of change of background slope is uniform over short intervals. (A warning is in order because the usual methods employ 'slope' to estimate the intensity at a second, unmeasured, offset position prior to calculating the background value by linear interpolation.)

As a rule of thumb, count on the specimen for the time it would take to record 10^8 counts on a pure standard. If the atomic numbers of trace element and matrix are not too different then, all being well, the variation in the background will give error estimates of the order of a few 10ppm as defined by the $+/-3\sqrt{(bkgd)}$ criterion.

However a single count is not the preferred route, e.g.. 10 counts for 1/10th the time is a good deal better. This will help identify spurious signals from external sources like fluorescent lights switching on and off and, even more importantly, allow the instrument to measure the beam current at more frequent intervals than with a single long count.

Real Backgrounds The previous paragraph deals with an 'ideal' background, real ones are not so straight-forward. Any number of problems can be encountered:-
i) interference by other x-ray lines generated in the sample, e.g. Kβ lines falling near the Kα being measured, or multi-order reflections such as the Cu-Kα(4) on P-Kα(1). Fortunately the large difference in energy in the latter case means that the proportional counter can filter out almost all of the unwanted Cu radiation. A blank reference is a good way of ascertaining that filtering is complete or finding out how much still gets through.
ii) background 'holes' caused by diffraction effects in the analyzing crystal (19). A notorious example of this

is the dip in the background adjacent to the Au-Lα line recorded with an LiF crystal. See Figure 12.
iii) x-ray absorption edges in the counter window or counter gas. The background behavior near U-Mα is maybe the worst example of this type of trouble, see Figure13.
iv) spectral reflection of soft x-rays from the face of the analysing crystal. This is very pronounced for O-Kα from a lead stearate pseudo-crystal and is made worse on older instruments by back-scattered electrons reaching the counter window. A small permanent magnet suitably placed will eliminate the electron problem. The synthetic multilayer (SML) crystals now available can completely avoid this problem.

Finally with well-calibrated spectrometers it is possible to read the peak count on one spectrometer and the background on a second one (20), thus eliminating beam current changes as a source for concern.

Blanks and low level reference materials are in obvious demand when the above difficulties are considered. For large routine projects it is well worth the trouble to find or manufacture reference standards that are very close to the unknowns. With this in mind, ion-implanted materials have been tried for low level concentration standards (12). These have already seen use in calibrating SIMS data. A combination of great interest to date has been Au in sulfides such as arsenopyrite. As a minimum, simple on-off readings can be employed, i.e. first read on areas containing a known amount of gold then on gold free areas to obtain a basic calibration curve.

One final remark on technique, this sort of measurement should not be started on the first morning back after a shut down. The microprobe should be properly warmed up so that stage and spectrometers have all reached thermal equilibrium prior to taking data.

Light Element Analyses

There have been considerable advances in this area of EPMA. The new $\phi(\rho z)$ methods are intrinsically capable of correcting data gathered in the long wavelength region, but is the data free of systematic errors?

B, C, O, N and F are the chief interests and originally all were difficult to even detect. However the advent of SML crystals has increased count rates by large factors. As a result recent attention has turned to the subtleties of the chemical wavelength shifts that arise in compounds. The most familiar is the chemical shift like that found on the Si-Kα line when going from silicon metal to silicate. Similar shifts are found with other elements in that part of the Periodic Table, due in part to a favourable combination of relative size of shift and diffracting crystal resolution. With light elements there is also a shape change. This was found by Solberg (21) who first proposed using the integrated peak areas as the true measure of the emission intensity instead of simple peak heights. A further development along these lines shows that in some cases deconvolution of main peaks into their individual component emission lines is necessary for really accurate work(19). The next step is to decide what are the appropriate absorption coefficients to employ. For example the Fe-Lα:Lβ band changes shape with electron beam energy, see Figure 14. It is severely affected by self-absorption and would in fact require a continuum of absorption coefficients for its proper treatment. As a practical matter it is possible, in some instances, to record spectra free from self-absorption features by working on evaporated thin films. In general, using standards that are close chemically to the sample in question is the easy way out for now.

EDXS

The above remarks apply to microprobes with their omnipresent wavelength spectrometers. For the SEM-EDXS combination found in many non-specialist laboratories, a somewhat different set of items must be looked at. Here EDXS stands for Energy Dispersive

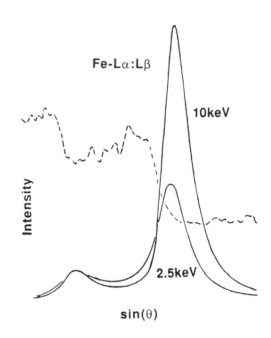

Fig. 14 Fe-Lα/Lβ spectra recorded at 2.5 and 10keV together with the experimentally determined x-ray absorption, (dotted line), as a function of sinΘ. The position of the absorption edge results in a self-absorption factor that varies strongly with wavelength thus distorting spectra at higher accelerating potentials.

X-ray Spectrometer / Spectroscopy. The much broader peaks found in the EDXS do not look too hopeful with regard to MDL's: just what is the best estimate of sensitivity? Peak channel over background channel, integrated peak against integrated background or integrated peak versus background channel? The last suggestion is, of course, a giveaway, it is found to be the best estimate of the MDL based on the above considerations (22). A moment's reflection will show how this is the nearest to the wavelength situation. Poor energy resolution in the EDXS spreads a given x-ray line into a band 150-200eV wide, many times its natural width of about 5eV. The true peak intensity is determined by the peak integral. For the background, things are a little different; whilst each narrow band of x-ray emission is spread out like the characteristic lines, it also receives compensating intensity from the adjacent wavelengths. As a result the single channel intensity is the best estimate for the background figure to use in the MDL formula. The problem of actually determining the peak and background levels still remains.

Acknowledgements

Whilst the discovery and continued development of the MSG$\phi(\rho z)$ theory at MTL and BRGM is in major part the work of the authors, they gratefully acknowledge the vital contributions made by others. In particular those of Keith Milliken(MTL), Wm Robinson (when at UWO) and J. D. Brown (UWO) for use of their $\phi(\rho z)$ experimental data base. Vera Weatherall ran most of our experiments and we thank M Phaneuf for use of the Ao P tracer. Figure 1 was run on the Electron Flight Simulator courtesy of Small World Inc, San Mateo, CA. The Galapagos chlorite is courtesy of the Geological Survey of Canada

Refences

1   Merlet, C. Mikrochimia Acta suppl 12,107- (1991)
2   Pouchou, J-L. and Pichoir, F. 10th Int. Cong. X-ray Optics Microanal., J. Physique, 45, C2-47 (1984)
3   Pouchou, J.-L., Anal. Chim. Acta 283, 81-97 (1993)
4   David C. Joy,"Monte Carlo Modelling for Electron Microscopy and Microanalysis ",Oxford University Press, Oxford,England (1995)
5   See for example Microbeam Analysis, Scanning or Scanning Microscopy.
6   Packwood, R.H. and Brown J.D., X-ray Spectrom. 10,138-146(1981)
7   S.J.B. Reed, "Electron Microprobe Analysis", Cambridge University Press, Cambridge, England(1975)
8   Cockett, G.H. and Davis, C.D. Brit. J.Appl. Phys. 14,813-816 (1963)
9   Packwood, R.H. and Remond. G., Scan. Microscy.6,367-384 (1992)
10   Packwood, R.H. and Brown J.D.Int.Chem. Cong. Pac. Basin Socs. (1984) unpublished proceedings
11   Martin, P., Hanford, M-A., Packwood, R., Dignard, L.and Moore, V. 2nd Int. Conf. Zn and Zn alloy Coat. Steel Sheet. Amsterdam. 112-116, (1992)
12   Remond,G., Packwood, R.H. and Gillies, C. Analyst 120, 47-1260 (1995)
13   By SAMx, 4,rue Galilée, 78280 Guyancourt - France
14   Waldo, R. A. Microbeam Anal. 310-314 (1988)
15   Bastin, G.F., Heijligers, H.J.M. and Dijkstra, J.M. microbeam Anal. 159-160 (1990)
16   Rickerby, D. G. and Thiot, J-F. Mikrochimica Acta 114/115,421-429 (1994)
17   Pozsgai. I. Euro. Microscy. Anal. 35,9-11 (1995)
18   Zou, H.,Hood, G. M., Roy, J.A.,Packwood, R.H. and Weatherall V. J. Nucl. Mater. 208,159- 168 (1994)
19   Remond,G., Campbell, J.L., Packwood, R.H. and Failin, M. Scan. Micrscoy. Supp.7,89-132 (1993)
20   McKay, G.A. and Seymour, R.S. Microbeam Anal. 431-434 (1982)
21   Solberg,T.N. Microbeam Anal. 148-150 (1984)
22   Geiss, R.H. and Savoy, R.J. Microbeam Anal. 59-61 (1991)

Appendix

The MSG$\phi(\rho z)$ theory assumes that the true $\phi(\rho z)$ can be modelled by curves of the form

$$\phi(\rho z) = \gamma_0 \exp(-\alpha^2(\rho z)^2) - (\gamma_0 - \phi_0).\exp(-\alpha^2(\rho z)^2 - \beta \rho z)$$

where

$$\alpha \approx 4.5 \times 10^5 \cdot \{(Z-N)/Z)\}.(Z/A)^{0.5}E_0^{-0.75}$$
$$\times [(Z/A)\log_e\{1.166.(E_0+E_C)/2J\}/(E_0^2 - E_C^2)]^{0.5}$$

Z is the atomic number, A the atomic weight, E_0 the accelerating potential and E_U the critical excitation potential, $N \approx 1.3$ and $J \approx 11.5 \times 10^{-3} Z$

$$\gamma_0 \approx 10\Pi.(U_0/(U_01)).\{1+ (10/\log_e U_0).(U_0^{-0.1}-1)\}$$

with the ionisation cross-section , Q, given by

$$Q.E_C^2 \propto (\log_e U_0/U_0^n, \quad U_0 = E_0/E_C \text{ and } n=0.9$$

For $\beta \approx 0.4\alpha Z^{0.6}$ and $\phi_0 \approx 1+2.8\eta(1- (0.9/U_0))$ where η is the electron backscatter coefficient.

To calculate the observed intensity, the $\phi(\rho z)$ equation is multiplied by the absorption factor, $\exp(-\chi\rho z)$, and then Integrated from zero to infinity:-

$$I_{obs} = C_A. \int_0^\infty \phi(\rho z).\exp(-\chi\rho z).d\rho z$$

where $\chi = \mu\cos\vartheta$, μ being the X-ray mass absorption coefficient and ϑ the X-ray take-off angle.
The terms γ_0 and α measure the height and width of the gaussian shape in the formula. ϕ_0 is the surface ionisation and β, together with γ_0 and ϕ_0 controls the slope of $\phi(\rho z)$ at the surface.
For a finite layer between depths δ' and δ we have the General Equation

$$I_{obs} = C_A(\sqrt{\Pi}/2\alpha).\exp(-\delta'(\chi'-\chi)).$$
$$[\gamma_0\exp(\chi/2\alpha)^2.\{Erf(\alpha\delta+(\chi/2\alpha))- Erf(\alpha\delta'+(\chi/2\alpha))\}$$
$$- (\gamma_0-\phi_0).\exp((\chi+\beta)/2\alpha))^2.$$
$$\{Erf(\alpha\delta+(\chi+\beta/2\alpha)) - Erf(\alpha\delta'+(\chi+\beta)/2\alpha))\}]$$

Where $\chi' = \mu'\cos\vartheta$ and μ' is the absorption in any over layer present.
For a uniform concentration specimen the intensity formula reduces to:-

$$I_{obs} = C_A(\sqrt{\Pi}/2\alpha).(\gamma_0\exp(\chi/2\alpha)^2.Erfc(\chi/2\alpha)$$
$$- (\gamma_0-\phi_0).\exp(\chi+\beta/2\alpha)^2.Erfc(\chi+\beta/2\alpha))$$

The various non-uniform concentration distributions can be handled in the following manner. For an exponential distribution centred at the surface of the form:-

$$\sigma_0 = \sigma_0\exp(- \kappa\rho z)$$

we make two changes. Here C_A becomes $\sigma_0.C_A$ and β transforms to $(\beta - \kappa)$. These result from mathematics - combining exponent parameters by adding them together.

High Spatial Resolution Chemical Analysis by TEM

R. W. Carpenter
Arizona State University, Tempe, AZ

Abstract

The nanoscale composition and structure of bulk three-dimensional solids, interfaces and thin films have important effects on the properties of materials, and are therefore of great interest. Transmission electron imaging and nanospectroscopy are the most well developed and familiar methods for these investigations. Several different resolution limits affect these measurements, particularly real space image resolution and spatial resolution and sensitivity of energy dispersive and electron energy loss spectroscopy. The current limits and very recent improvements for these parameters are discussed here. Relatively new chemically sensitive imaging methods, such as energy selected TEM imaging and Z-contrast imaging are also proving to be useful. Examples of applications are provided and discussed.

ANALYSIS OF SMALL REGIONS in solids requires determination of chemical distributions near defects or interfaces, and images to determine the geometric positions of atoms in the material. Typical lengths or "resolutions" associated with these measurements range from several hundred to one or two Angstroms. Various electron microscopy methods are required to attain the necessary spatial or image resolution for these observations.

Relative crystallographic misorientations across a crystalline interface or between a small particle and its surrounding matrix can be determined to within a small fraction of a degree by careful electron diffraction experiments. These methods are well developed. Spatial resolutions of 1nm and angular resolutions of 0.1mr are attainable(5). The most familiar method for determining chemical or elemental distributions is small focused probe x-ray or electron energy loss nanospectroscopy. Here, resolution means spatial resolution, which is equated with the size of the incident focused probe in the thin specimen limit, and

depends on the spatial position temporal stability of the probe during spectrum acquisition. Newer, less familiar methods for determining the same information are energy selected transmission electron microscopy imaging (ESTEMI) and Z-contrast or hollow cone imaging. The first of these is a direct chemical imaging method, that is, it is spectroscopic and identifies specific elemental distributions. Z-contrast and hollow cone imaging depend on spatial variations in atomic number in the specimen for image contrast and are therefore indirect methods for determining chemical distributions. Resolution considerations for these methods are somewhat different than for the small probe methods. The imaging methods used to determine displacement fields and atom positions are the familiar diffraction or amplitude contrast and high resolution imaging (HRTEM). Brief comments on the current state of the imaging methods are given, and, more importantly, on their relationship to the methods for determining chemical distributions.

The various methods are illustrated using applications to interfaces in edge-on orientation, because the structural width or geometric sharpness of these microstructural features is clearly visible in images. Physical intuition allows the reader to visualize what the expected chemical distributions will be, and to appreciate the most useful attributes of the chemical distribution measurement methods.

Resolution Limits

The most familiar resolution limit is the Scherzer or so-called interpretable resolution limit, δ, for HTREM imaging. It corresponds to spacing (i.e., length) in the specimen defined by the first zero of the contrast transfer function, which is dependent on objective lens defocus $D = \Delta f(C_s \lambda)^{-1/2}$, where D is the generalized defocus, Δf is the objective lens defocus, C_s is the objective lens spherical aberration constant, and λ is the incident electron wavelength(1). This resolution limit $\delta = 0.7C_s^{1/4}\lambda^{3/4}$ at D=1, and corresponds to spacings in the object (the specimen) imaged without phase reversals of the scattered amplitudes caused by the objective lens, so image detail at or above this resolution limit can be interpreted directly. The most recent TEMs, with maximum operating voltages in the 200 to 1000kV range, have interpretable resolution limits in the range of 1 to 2 Angstrom range(2). This resolution is sufficient for determination of the structure of grain boundaries and interfaces in terms of the distribution of columns of atoms when the boundaries or interfaces are viewed in edge-on projection, even for small unit cell crystals. These experimental methods have been used, for example, to show that the structure of a $\Sigma=13(510)$ [001] tilt grain boundary in silicon was in good agreement with the CSL model, with a periodic repeat distance of 13.9 Å (3).

The corresponding resolution limit for local chemical analysis by nanospectroscopic methods is the spatial resolution limit, given in the simplest approximation for the thin specimen case by the size of the small focused electron probe incident on the specimen. In this context the microscope pre-specimen lenses are used to form a small focused probe on the specimen. The probe is an image of the electron source. The alignment of the probe forming lenses and their aberration constants and defocus thus have important effects on attainable spatial resolution for nanospectroscopy. Misalignment can result in order of magnitude or more loss of spatial resolution. The loss is usually anisotropic in the object plane, where the probe is not circular. Alignment can be examined and adjusted in real time by directly observing a real space image of the probe in analytical microscopes of the TEM/STEM type that contain post specimen lenses(4,5). However, additional considerations also affect this resolution limit. Nanospectroscopy is count rate limited, so it is necessary to consider current in the probe as a function of probe size and the electron optics of the probe forming system. For a correctly aligned illumination system the total probe size is given to a good approximation by an incoherent (quadrature) sum of coherent broadening mechanisms(5):

$$d_t^2 = d_g^2 + d_f^2 + d_s^2, \text{ where} \qquad (1)$$

$$d_g = \frac{2}{\pi\alpha}\left(\frac{i_b}{\beta}\right)^{1/2}; \quad d_f = (0.61)\frac{\lambda}{\alpha}; \quad d_s = (0.3)C_s\alpha^3 (2)$$

Here, d_g is size of a Gaussian probe with current i_b, from a source of brightness β. The convergence half-angle of the probe is α. The other two terms, d_f, which is the probe broadening by diffraction from the edges of the probe defining aperture, and, d_s, which is the broadening resulting from spherical aberration of the probe forming lens, represent electron optical effects from the illumination system optics that cause the total probe diameter, d_t, to be larger than the Gaussian probe diameter, d_g. For experiments that are not count rate limited (i.e. not probe current limited) such as nanodiffraction or, in most cases, scanning imaging, d_g is neglected. A point electron source is assumed and the spatial resolution is then determined exclusively by d_f and d_s which depend only electron wavelength, divergence and spherical aberration. However, when probe current is important, such as in nanospectroscopy, d_g must be included, and the brightness of the electron source, β, becomes important. Brightness varies dramatically with the type of electron source used, from about 5×10^8 amps $(cm^2.steradian)^{-1}$ for field emission sources down to 10^5 for tungsten thermal emission sources. For a given probe current, d_g will be much smaller when high brightness sources are used. Source brightness is the primary reason that all new high performance analytical electron microscopes use field emission sources. The energy spread of electron beams emitted by field emission sources is also small (<0.5eV) compared to thermionic source energy spread (>2.5eV), so that probe broadening caused by chromatic aberration in the microscope illumination system can usually be neglected when field emission sources are used, as it is in the equation for d_t above.

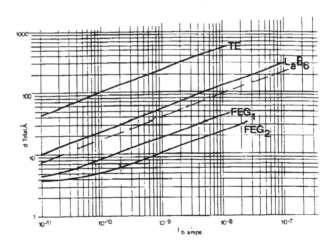

Fig. 1. Probe size as a function of probe current for sources of different brightness β in the incoherent approximation. Aperture and spherical aberration broadening are included, 100 kV, C_s=1.5 mm. β (A cm^{-2} sr^{-1}): TE=10^5; LaB$_6$=5×10^6; FEG$_1$=10^8; FEG$_2$=5×10^8.

Some example calculations of probe diameter, d_t, as a function of probe current, i_b, for sources of different brightness using the optimum values of α are shown in Figure 1. They show that field emission sources are necessary to obtain probes smaller than 10Å containing 0.1 to 1 nanoamp. It is also well known that field emission source probes have higher partial coherence than thermionic probes. Calculations assuming both full coherence and incoherence have shown that the incoherent assumption is appropriate for field emission probe size calculation when realistic divergence half angles, α, are used (1 to ~15 milliradians)(4).

The net count rate into an electron energy loss spectrum (ELS) edge is given by

$$R = i_b \, \sigma \, t \, N \, \Delta E, \text{ where} \quad (3)$$

i_b is the focused probe current (electrons sec^{-1}) as before; σ is the differential inelastic scattering cross section ($nm^2 \cdot atom^{-1} \cdot eV^{-1}$); t is the specimen thickness (nm); N is the target atom density (nm^{-3}), typically 50 for materials like SiO_2 or SiC; ΔE=channel width or energy spread (eV), typically 1eV for a field emission source. A similar count rate equation is applicable to energy dispersive x-ray spectroscopy(6). The number of counts desirable in an absorption edge depends on the experimental objective; 5000 is, conservatively, often enough for quantitative elemental analysis but more may be required for observations of near edge fine structure bonding effects or impurity detection. During the time the spectrum is being acquired, the positional stability of the specimen and probe will affect the attainable spatial resolution. Movement of the probe caused by electrical instabilities can degrade the spatial resolution to well above the incident probe size. Short acquisition times are desirable, consistent with the electron irradiation effects tolerance of the specimen. Measured values of σ (10mr ELS collection half-angle) at core edges of interest such as silicon L or oxygen, nitrogen or carbon K edges in the corresponding compounds with silicon range from 10^{-7} to 5×10^{-9} $nm^2 atom^{-1} eV^{-1}$(7). The specimen thickness, t, must be restricted to small values, ~10nm, to avoid undesirable multiple scattering effects. This small thickness does not present unusually severe experimental difficulties. It is the same thickness required for interpretable HRTEM images. All parameters in the count rate equation are now fixed except the probe current. It was noted above that attainable currents in field emission probes on the order of 1nm in diameter are in the range of 0.1 to 1.0 nanoamp. Then the maximum count rate (for $\sigma = 10^{-7}$ and i_b=1nA) is about 3×10^5 counts s^{-1} and the minimum value of R (for $\sigma = 5 \times 10^{-9}$ and i_b = 0.1 nA) is about 1600 counts s^{-1}. If spectra are acquired containing ~10,000 counts in edges of interest, such as the Si-L or O-K edges, the time for acquisition will vary between about 30 ms and 6 s per *channel* at the peak position. Reductions in spectrum acquisition time using recently developed parallel collection hardware/ software have enabled dramatic improvements in spatial resolution for electron energy loss spectra(8). Spectra are normally recorded in 1024 channels. Using parallel acquisition a spectrum of, say, 1000eV width can be recorded in 30 ms to 6 s, depending on experimental details. Using older serial acquisition technology, recording the same spectrum would require a 1024 fold increase in acquisition time, i.e. 30 s to 100 min. During aquisition spatial resolution will degrade if the focused probe moves relative to the specimen or conversely, so very stable beam steering electronics and mechanical specimen stages in

analytical TEM's are essential in addition to fast parallel EELS collection to obtain the best spatial resolution. The most recent field emission analytical TEM I have examined will maintain relative focused probe/specimen position within 1nm per minute, which will lead to an uncertainty in position of 0.1nm, or 10% of the probe diameter when a 1nm probe is used. Longer collection times, such as required for energy dispersive x-ray spectroscopy, will decrease attainable spatial resolution. From the practical experimental viewpoint, a spatial resolution of 2nm can be routinely achieved using thin specimens if the conditions above are met. Under the best experimental conditions, with sufficient effort, I expect that spatial resolution better than 1nm can be achieved using the most recent instrumentation and methods.

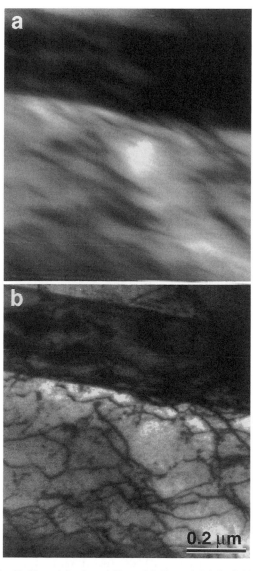

Fig. 2. Unfiltered (top) and filtered (ΔE=4ev) bright field diffraction contrast images of 208 nm thick austenitic stainless steel containing dislocations and thin twins (120 kV). Thickness measured from projected twin width. The resolution improvement by filtering is equivalent to examination in a high voltage TEM. Foil normal and B = [011], g = <111>. Ω Zeiss 912 microscope.

Energy selected or energy filtered imaging can be accomplished either by using an Ω filter (which is a type of electron spectrometer) on the electron exit side of the specimen in the optical path of the TEM column or an imaging energy filter at the end of the optical path, after the projector lens(9,10). Both of these instruments, which are just beginning to be applied to problems in materials science, allow image formation with transmitted electrons of a selected energy bandwidth. A bright field *filtered* image is formed by imposing a narrow energy band window about the forward scattered beam, removing low-loss (plasmon and valence band) and core-loss inelastically scattered electrons from the image. A striking example is shown in Figure 2. The exclusion of inelastic scattering from the image of stainless steel removes chromatic blurring and is equivalent to imaging the specimen in a high voltage electron microscope without the attendant displacement irradiation effects. An *energy selected* image is formed by aligning the inelastically scattered electrons corresponding to the element one wishes to detect along the optic axis of the microscope and forming an image with them. The alignment with the optic axis is made by changing (increasing) the accelerating voltage an amount equal to the energy loss of interest. When core loss electrons are used, these images have direct chemical sensitivity. Where the image is bright the concentrations of the element is high and conversely. The resolution of these chemical images is reduced by chromatic aberration corresponding to the width of the energy window, typically 20eV, that brackets the core loss energy of interest. This resolution is $\delta = C_c(\Delta E/E)\alpha$, where C_c is the chromatic aberration coefficient of the microscope, ΔE is the window width, E is the accelerating voltage and α is the divergence half-angle. For $C_c = 2.7mm$, $\Delta E = 20eV$, $E = 120keV$, and $\alpha = 2.5$ milliradian, $\delta =10Å$. Practical considerations such as mechanical stage stability during recording of energy selected images also affect resolution in the same way that spatial resolution of spectra are affected. These images are usually recorded in parallel using a CCD camera, but the intensity is low because inelastic scattering cross sections are relatively small. Increasing α to increase the incident current can reduce exposure time, at the expense of increasing δ. The practical limits on these parameters, i.e. experimental techniques, have not yet been fully explored.

Z-contrast imaging, which is sometimes called high angle annular dark field scanning transmission imaging, is sensitive to the atomic number, Z, of the specimen. Heavy elements scatter electrons (elastically) out to higher angles than lighter low-Z elements. Hence a localized region containing a higher than average concentration of heavy elements will appear brighter than its surroundings in a Z-contrast image. This imaging technique is indirectly chemically sensitive, since it can locate heavy-element-rich regions, but it does not give the local value of Z. Complementary spectroscopy must be used to identify the heavy element chemistry. The electron optical details of Z-contrast imaging have been discussed(11,12). Z-contrast is a high resolution imaging method; resolutions of ~1.5Å have been achieved. Hollow-cone imaging is the TEM analog of Z-contrast STEM imaging, and will produce the same image resolution in a TEM with suitable electron optical and stability parameters(13).

Applications to Materials Research

We will use some applications to chemical distributions at interphase interfaces and matrix grain boundaries in a ceramic matrix composite (CMC) to illustrate these methods. This type of composite is being developed to alleviate the brittleness problem of monolithic ceramics at lower temperatures where dislocations are not mobile, by providing high energy absorption paths for crack propagation in the CMC microstructure. CMCs are typically synthesized by mixing the particulate starting materials with a small amount of sintering aid, followed by processing at high temperature to achieve theoretical density. The sintering aids are chosen to form a small volume fraction of liquid with some of the starting material, so that densification proceeds rapidly by liquid phase sintering, rather than very slowly by solid state mass transport in these covalent solids. The small volume friction of liquid often results in formation of a thin amorphous film or phase along interfaces and boundaries upon cooling. The chemistry and distribution of this thin amorphous phase, and its interactions with the crystals bounding it are the primary interests. It is thought to be the preferred crack path during low temperature fracture. At high temperature it can undergo viscous flow under stress, resulting in creep deformation. The rate should be chemistry dependent. Many examples of these phases have been examined by HRTEM imaging, but very few direct chemical measurements have been reported.

The HRTEM image of figure 3a shows an edge-on matrix grain boundary in an β-Si_3N_4 CMC reinforced with 20 vol.% SiC whiskers. A thin (~0.8 nm) amorphous phase is clearly visible. The CMC was synthesized by mixing the nitride powder and whisker starting materials with 5.5 wt.% Y_2O_3 and 1.1 wt. Al_2O_3, pressing and pre-sintering for 1 hr. at 1500°C in 0.1 MPa argon. Complete densification was achieved by liquid phase sintering during hot-isostatic-pressing at 1780°C in 190 MPa argon for 1 hr. The oxygen and nitrogen distributions at the grain boundary are shown by the position-resolved electron energy loss (PREELS) plots shown in figs. 3b, c, respectively. The boundary region is enriched in oxygen and depleted in nitrogen, which, to first approximation, is what one would expect for a non-oxide CMC synthesized with oxide sintering aids. The PREELS plots were made by stepping a 3 nm diameter focused probe along the

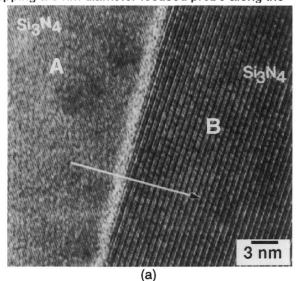

(a)

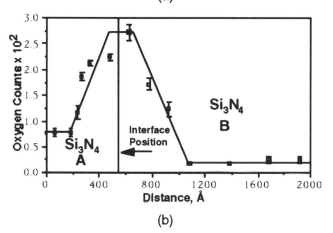

(b)

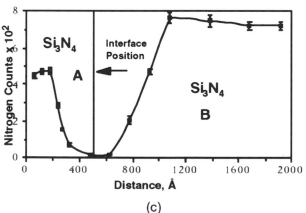

(c)

Fig. 3. (a) HRTEM image of β-Si_3N_4 grain boundary with thin amorphous phase ~ 0.8 nm thick. (b) Variation in oxygen content across the boundary shown in (a), along the arrowed scan path. (c) Variation of nitrogen content across the same scan path. The chemical width for both oxygen and nitrogen is ~87 nm.

path indicated by the arrow shown crossing the edge-on boundary in fig. 3a. Spectra were acquired as described above at each step, and for each spectrum the net counts were calculated in the appropriate edge

and plotted as a function of step position relative to the edge-on boundary. Details of the computer equipment for performing PREELS experiments have been given (14). Notice that the width of the oxygen and nitrogen distributions is about 80 nm. These distribution curves are convolutions of the current distribution in the focused probe and the actual chemical distribution, i.e., they are broadened by the probe current distribution. In other experiments we have shown that the probe broadening is about 6 nm (15). This correction is small (~8%) relative to the width of the oxygen and nitrogen distributions. Because the oxygen and nitrogen distributions are wider than the thickness of the amorphous layer visible in the corresponding HRTEM image, two widths are required to describe this boundary: a *chemical width* from the oxygen and nitrogen distributions and a *structural width* from the HRTEM image. For this boundary the ratio of chemical to structural width is about 50. The structural width of matrix grain boundaries in these CMCs varies from one boundary to another, possibly because the mixing operation during synthesis resulted in a non uniform sintering aid distribution. Figure 4a shows a grain boundary with small structural width, ~0.5 nm. This structural width is only slightly larger than the SiX_4 tetrahedral structural units (~0.3 nm) present in tetrahedrally bonded silicon compounds and glass, so it cannot easily be described as a thin amorphous second phase, but it is a region of structural disorder between the two grains. Nevertheless the PREELS oxygen and nitrogen distributions of figs. 4b and 4c show that chemical widths ~40 nm wide are present at this boundary. When these are corrected for probe broadening, the ratio of chemical to structural widths is ~70, about the same as the wider boundary shown in figure 3, although both the structural and chemical widths are smaller. By comparing the HRTEM images of the two boundaries, figs. 3a and 4a, it can be seen that images are not a reliable indicator of the presence of chemical widths. The wide boundary image (fig. 3a) is a good qualitative indicator for the presence of chemical impurities but the narrow boundary image (fig. 4a) is not.

The distributions of the other elements expected in grain boundary regions on the basis of sintering aid composition are also of interest. Diffusion of oxygen ions into the Si_3N_4 crystal regions and depletion of nitrogen ions from the same regions would create a charge imbalance unless compensated by substitution of aluminum and yttrium for an equivalent amount of silicon in the same regions. The distribution of yttrium, with atomic number 39, which is appreciably heavier that the other constituents whose average atomic number is 10, can be examined by Z-contrast imaging, provided that an appropriate spectroscopy is used to establish that the origin of the expected high contrast in the grain boundary region is indeed yttrium. Figure 5a is a plot of Z-contrast image intensity across a matrix grain boundary in edge-on orientation, made by scanning a focused probe along a line perpendicular to the boundary, similar to the scan paths indicated by arrows in figures 3a and 4a. Note that the contrast

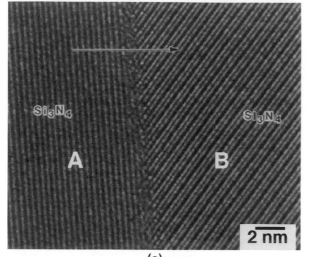

(a)

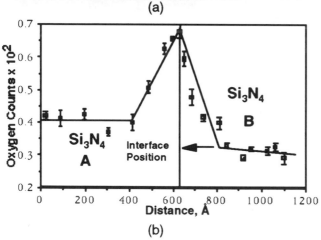

(b)

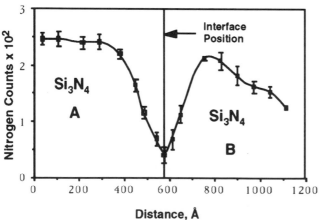

Distance, Å

(c)

Fig. 4. (a) HREM image of an edge-on matrix grain boundary in a Huber SiC(w) composite. The width of the disordered region at the boundary, i.e., the structural width, is about 0.3 nm. Scan path is marked by the arrow. Grains A and B are oriented for g = <110> and <100> two beam diffraction. (b) Variation in oxygen content with distance from the boundary for the scan path shown in Fig. 4(a). The chemical width of this oxygen distribution was about 41 nm. (c) Variation of nitrogen content with distance from the boundary for the scan path shown in Fig. 4(a). The chemical width of this nitrogen distribution at the interface was about 45 nm.

width is ~20 nm, much greater than the 0.5 to ~1.0 nm structural widths observed using HRTEM. Windowless energy dispersive x-ray spectra (light element sensitive) recorded from the boundary itself (solid curve) and 14 nm away from the boundary (dashed curve) are shown in figure 5b. Yttrium, aluminum, silicon and nitrogen are present in the boundary and silicon and nitrogen are easily detected at a point 14 nm away. When the dashed 14 nm spectrum is amplified by a factor of 10 (light dashed curve) shoulders on the high energy side of the nitrogen-K peak and on both sides of the silicon-K peak indicate clearly that aluminum, yttrium and oxygen are present 14 nm away from the boundary plane. Note that the relative abundance of aluminum is now larger than yttrium, which is the reverse of the situation at the grain boundary. It is unusual for yttrium to dissolve in β-Si_3N_4, however our result is in agreement with x-ray diffraction lattice parameter measurements that indicated yttrium dissolution does occur in β-Si_3N_4 in the presence of aluminum (16). Figure 5c shows an oblique view of the Y-modulated Z-contrast image of this grain boundary, indicating that the chemical width is continuous laterally along the boundary. These results show that the aluminum and yttrium distributions follow the oxygen distribution.

The chemical widths, i.e. distributions, observed using the same small probe methods, at SiC whisker/Si_3N_4 matrix interfaces were qualitatively the same, but with two significant differences. First, the chemical widths were laterally discontinuous along the interfaces, and second, where chemical widths existed they tended to be asymmetric with respect to the interface plane, extending farther into the nitride than into the carbide. The discontinuous character is related to the known surface faceting of SiC whiskers which resulted in regions of near point contact with Si_3N_4 crystals, and the chemical width asymmetry is related to lower solute solubility in the carbide phase (17).

Energy selected transmission electron imaging (ESTEMI) was used to make real-space maps of the chemical widths of most of the sintering aid constituents that we have so far examined at specific positions on a particular boundary by PREELS or conventional point spectra. These images can be simply visualized as dark field images made with inelastically scattered electrons whose energy corresponds to a core edge of an element of interest, i.e. aluminum-L, oxygen-K, nitrogen-K and carbon-K for this CMC. Some results are shown in figure 6. These images were recorded digitally using a slow scan CCD camera and have been processed to remove pre-edge background intensity, as described earlier (18). The images show a long section (~155 nm) of matrix grain boundary connecting to triple junctions in the CMC. Figure 6a is conventional bright field image, and figures 6b-6f are spectroscopic chemically sensitive ESTEMI, where bright (above

corresponds to a higher concentration of the element

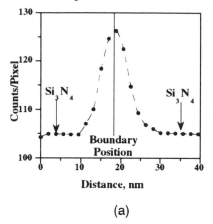

(a)

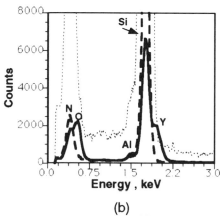

(b)

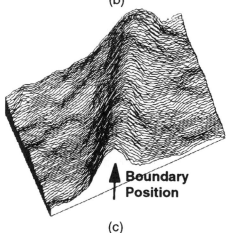

(c)

Fig. 5. (a) Z-contrast image intensity line scan across an edge-on matrix grain boundary in the Huber composite. The intensity is proportional to yttrium content. The boundary is a mirror plane for the yttrium distribution curve. Other solutes are present, as shown in Fig. 5(b) (solid curve) and displaced 14 nm away from the boundary plane into one of the bounding crystals (dashed curve). The latter spectrum is also shown amplified by 10x (narrow dotted curve). The asymmetric shoulders on the Si and N peaks in the amplified spectrum show the O, Y and Al are present in the matrix 14 nm away from the boundary plane. Note that more Al than Y was present in the matrix, the converse of these element contents in the plane of the boundary. (c) Oblique view of amplitude-modulated Z-contrast image of this matrix grain boundary in the edge-on orientation. The yttrium distribution is laterally continuous along the boundary and

there is no evidence for differential ion thinning at this particular boundary of interest than in the adjoining crystals. Figures 6b-6e show that the grain boundary region is rich in aluminum, poor in silicon, rich in oxygen and poor in nitrogen, in agreement with the previous PREELS results, but over a much larger lateral region of boundary. Image recording time was up to 30 seconds. The last image, figure 6f, was recorded using the carbon-K edge and is sensitive to contamination, but it also includes the weak yttrium M3 (300 eV) and M2 (312 eV) edges, which is the probable reason for the bright triple junction contrast. The results are in very good qualitative agreement with the earlier PREELS results. The graphs below each image are plots of the image intensity along a strip 50 pixels wide (white box in Fig. 6a) normal to the edge-on boundary, and provide information similar to a PREELS plot. The corresponding structural width (fig. 6a) and chemical widths (figs. 6b-6e) are shown by vertical lines and numerically on each graph. The signal to noise ratio of the boundary in fig. 6f was too low to provide useful information. For three different boundaries examined, the ratios of chemical to structural width ranged from 1.2 to 2.8. The larger chemical widths are again in excellent qualitative agreement with the PREELS results, but the absolute value of these chemical widths is appreciably lower than the PREELS values above. A primary reason for the quantitative disagreement in widths is the differences in incident current density for ESTEMI and PREELS measurements. The incident probe current density for PREELS was $\sim 10^4$ amps/cm^2 (Philips field emission 400 ST microscope), but for ESTEMI the incident image current density was about 5 amps/cm^2 (Zeiss912 thermionic emission microscope) (19). This difference has an important effect on ability to detect small solute concentrations, since the minimum detectable mass fraction in an EELS experiment is proportional to $(Jt)^{-1/2}$, where J is the incident electron flux and t is the acquisition time (20). The difference in flux indicates that PREELS is ~50 times more sensitive for detection of small solute concentrations than ESTEMI, assuming other experimental variable of the two methods are equal, which resulted in loss of the ability to detect the low concentrations of sintering aid elements in the extended tails of the chemical width using ESTEMI. The difference in detection sensitivity was partially compensated by extending the image acquisition time for ESTEMI, but stage drift precluded times longer than ~30 seconds. Despite this difference in sensitivity, this new method of *chemical microstructure analysis* is expected to find widespread acceptance among materials scientists. It is quick and convenient, and will yield immediate physical insight into many chemical distribution problems.

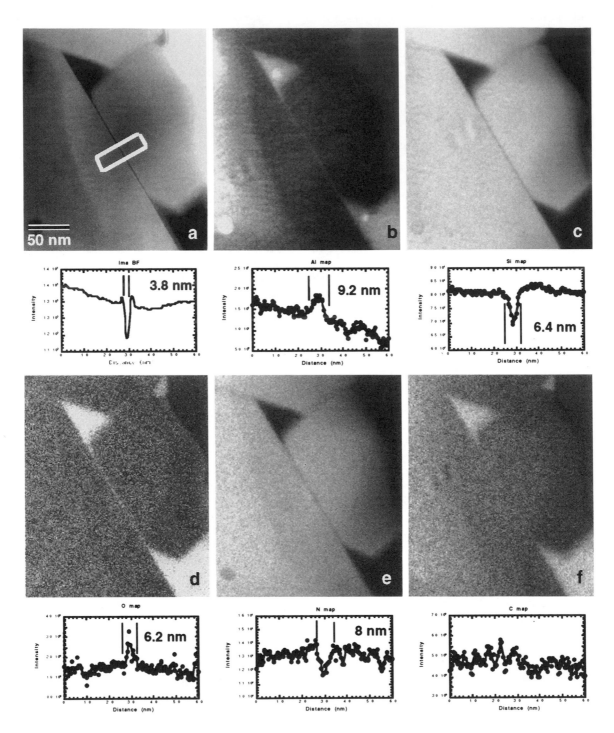

Fig. 6. Bright field and energy selected images of an edge-on grain boundary. (a) Bright field; (b) Al-L, energy window (EW) = [80,100], 10s collection time (CT); (c) Si-L, EW = [120, 140], 10s CT; (d) O-K, EW = [545, 565], 30s CT; (e) N-K, EW [425, 445], 30s CT; (f) C-K, EW = [300, 320]. 20s CT.

Summary and Conclusions

The PREELS, Z-contrast, point spectra and HRTEM results discussed here established that chemical widths formed during processing Si_3N_4 matrix CMCs reinforced with 20 vol.% SiC whiskers and liquid phase densified by hot isostatic pressing at 1780°C for 1 h with Y_2O_3 + Al_2O_3 sintering aids were 10 to 120 times wider than the grain boundary and matrix/whisker interface structural widths determined by HRTEM. Further, the chemical widths were formed by in-diffusion of sintering aid elements into the bounding crystals, and were laterally continuous along matrix grain boundaries but laterally discontinuous along matrix/whisker interfaces (17,21,22). The chemical widths correspond to solute diffusion constants of ~3 x 10^{-15} to 6 x 10^{-16} cm^2/s, characteristic of slow diffusion in covalent crystals.

Four different microscopes were used to make the measurements including the first evaluation of ESTEMI for CMC interface/boundary research: a Philips 400 field emission AEM for spectroscopy, a TOPCON 002B for HRTEM, a vacuum generators HB-501 for Z-contrast and a Zeiss 912 Ω for ESTEMI. This was a rather complicated experimental procedure, particularly the process of finding the same small microstructural area in one specimen in several different microscopes. Microscopes now becoming available, when coupled with imaging energy filters, PREELS and x-ray instrumentation and digital data recording devices, promise major improvements for this type of materials microstructure research. The basic microscope required is an HRTEM operating at a maximum voltage of 200 kV with a field emission electron source, to enable focused probe formation at the specimen with probe diameter down to or below 1nm and with probe current in the 0.1 to 1 nanoamp range. Several microscopes of this type have been described (2). The 200 kV limit on operating voltage is important to eliminate or minimize knock-on irradiation effects in most materials of interest. It is important that the specifications for a microscope to be used for high resolution analysis of chemical distributions and structure include limits on stability of focused probe position on the specimen, as well as the usual image resolution specification. The former limits will require both mechanical stability of the stage and electrical stability of the probe position control electronics. The newest microscope I have examined will maintain probe/specimen position within 1 nm per minute, along with image resolution better than 0.2 nm using a field emission source, which provides the required current in a small probe. We are presently evaluating all of the characterization methods that can be performed with satisfactory resolution in this microscope, which formerly required at least four separate microscopes to accomplish.

Acknowledgment

The research discussed in this paper is supported by the U.S. Department of Energy, Basic Energy Sciences, Division of Material Sciences under Grant No. DE-FG03-94ER45510 (Dr. O. Buck).

References

1. J.C.H. Spence, "Experimental High-Resolution Electron Microscopy", Clarendon Press, Oxford (1981).
2. O'Keefe, M.A., Ultramicroscopy, 47, 282-297 (1992).
3. Kim, M.J.; Carpenter, R.W.; Chen, Y.L.; and Schwuttke, G.H., Ultramicroscopy, 40, 258-264 (1992).
4. Weiss, J.K.; Carpenter, R.W.; Higgs, A.A.; Ultramicroscopy. 36, 319-329 (1991).
5. Carpenter, R.W. and Spence, J.C.H., Jour. of Micros. 136 pt. 2, 165-178 (1984).
6. N.J. Zaluzec, "Quantitative X-ray Microanalysis: Instrumental Considerations and Applications to Materials Science", p. 121 in Introduction to Analytical Electron Microscopy, ed. by J.J. Hren et. al., Plenum Press, New York (1979).
7. Skiff, W.M.; Carpenter, R.W. and Lin, S.H., J. Appl. Phys., 62, 2439-2449 (1987).
8. Krivanek, O.L.; Ahn, C.C. and Keeney, R.B.; Ultramicroscopy, 22, 103-116 (1987).
9. Kohl, H. and Rose, H., Advances in Electronics and Electron Physics, vol. 65, 173-227 (1985).
10. Krivanek, O.L.; Gubbens, A.J. and Delby, N., Micros. Microanal. Microstruct., 2, 315-332 (1991).
11. Jesson, D.E. and Pennycook, S.J., Proc. R. Soc. Lond. A 441, 261-281 (1993).
12. Liu, J. and Cowley, J.M., Ultramicroscopy 37, 50 71, (1991).
13. Dinges, C.; Kohl, H. and Rose, H., Ultramicroscopy 55, 91-100 (1994).
14. Weiss, J.K.; Rez, P. and Higgs, A., Ultramicroscopy, 41, 291-301 (1992).
15. Catalano, M.; Kim, M.J.; Carpenter, R.W.; Das Chowdhury, K. and Wong, J., J. Mater. Res., 8, 2893-2901 (1993).
16. Loehman, R.E. and Rowcliffe, D.J., J. Amer. Ceram. Soc., 63, 144-148 (1980).
17. Das Chowdhury, K.; Carpenter, R.W.; Braue, W.; Liu, J. and Ma, H., J. Amer. Ceram. Soc, 78, 2579-2592 (1995).
18. Carpenter, R.W.; Bow, J.S.; Kim, M.J.; Das Chowdhury, K. and Braue, W., Mater. Res. Soc. Sympos. Proc. Vol. 357, 271-276, Mater. Res. Soc., Pittsburgh (1995).
19. Berger, A.; Mayer, J. and Kohl, H., Ultramicroscopy, 55, 101 (1994).
20. Isaacson, M. and Johnson, D., Ultramicroscopy, 1, 33 (1975).
21. Das Chowdhury, K.; Carpenter, R.W. and Braue, W., Ultramicroscopy, 40, 229-239 (1992).
22. Das Chowdhury, K.; Carpenter, R.W. and Braue, W., Mater. Res. Soc. Sympos. Proc. Vol. 238, 421-426, Mater. Res. Soc., Pittsburgh (1992).

Advanced Materials Characterization Using AEM and APFIM

M. G. Burke

Westinghouse - Bettis Atomic Power Laboratory, West Mifflin, PA

MATERIALS DEVELOPMENT REQUIRES the evaluation and characterization of structure and properties for optimization of performance in engineering applications. Both high performance materials and more conventional alloys can be tailored for specific applications by careful microstructural control. This is generally accomplished via thermomechanical processing or thermal treatment. To obtain the necessary correlations between processing, properties, and structure, thorough microstructural characterization is frequently required. In this paper, the application of two advanced microstructural characterization techniques, analytical electron microscopy (AEM) and atom probe field-ion microscopy (APFIM) to the analysis of metals and alloys will be described. An overview of both techniques, including recent advances in APFIM analysis, will also be presented.

Analytical Techniques

AEM. Analytical electron microscopy is a widely used technique for the characterization of materials because it has the capability to provide both microstructural and microchemical data for a broad spectrum of metals, ceramics, composites, etc. The AEM incorporates a conventional transmission electron microscope (TEM) with scanning transmission electron detectors, and the capability for microanalysis such as an energy dispersive x-ray spectrometer and/or electron energy loss spectrometer. Frequently, these instruments are equipped with secondary and backscattered electron detectors, so that an extensive range of imaging and diffraction analyses are possible. In addition to its operation as a transmission electron microscope for general morphological and conventional electron diffraction analyses, the AEM can be operated in the scanning transmission mode (STEM) and thus produce a fine (~ 5 to 10 nm and larger) electron probe which can be employed for microdiffraction and energy-dispersive x-ray spectroscopy (EDS) microanalysis. The spatial resolution of the conventional AEM for microchemical analysis is generally >~ 20 nm. The field emission gun AEM, however, has a greatly improved spatial resolution due to the very fine (1-2 nm) probe size and high current density which is attainable. The capabilities of AEM and developments in the technique are discussed in detail by Williams (1) and in this symposium.

APFIM. Atom probe field-ion microscopy (APFIM), originally developed by Müller and Panitz in 1967 (2), is another analytical technique for the near-atomic level characterization of materials. Although the applications of this technique have not been as extensive as AEM, APFIM has unique capabilities which complement and extend the microstructural data obtainable by AEM. The APFIM consists of a field-ion microscope (FIM) coupled with a time-of-flight mass spectrometer, shown

schematically in Figure 1. Materials which can be analyzed by the conventional APFIM technique must be electrically conductive. The APFIM specimen is a fine needle which has a tip radius < ~50 nm. The needle specimen is inserted into the field-ion microscope which operates under ultrahigh vacuum. A cryogenic cooling system maintains the needle specimen at a temperature between 20 and 100 K (-253 and -173°C). The specific temperature is selected based upon the nature of the material and the experiment to be conducted. A small amount of inert gas (generally Ne) is admitted into the system to provide the imaging medium. The microstructural examination is performed via the field-ion image which is obtained by increasing the positive voltage on the tip until field ionization of the imaging gas occurs. As the gas atoms are ionized at the tip, they are projected radially away from the tip towards the electron channel plate-phosphor imaging screen assembly where the field-ion image is formed.

As the tip voltage is further increased, the surface atoms in ledge positions become ionized and follow a similar trajectory to the imaging assembly which consists of an electron channel plate for image intensification located in front of a phosphor screen. A small hole in the microchannel plate/phosphor screen imaging assembly serves as the entrance aperture to the time-of-flight mass spectrometer. A single ion detector is located at the end of the flight path. The atom probe also requires a high speed timing system for measurement of the flight times of the ions as they leave the tip and strike the single ion detector. The microchemical analysis is performed by computer-controlled pulsed field-evaporation coupled with the measurement of the flight time. The evaporation voltage and flight time of each ion reaching the single ion detector are used to calculate the mass-to-charge ratio (m/c) of each ion. The m/c for each analyzed ion is then stored on the computer in the sequence of arrival at the detector. The chemical identity of each ion is determined by its m/c. The calculation of composition is very straightforward for atom probe data; to determine the atomic percentage of element A in a material, simply divide the number of A ions detected by the total number of ions collected. Atom probe analysis requires careful

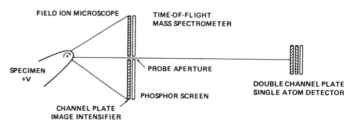

Figure 1. Simplified schematic diagram of a straight time-of-flight atom probe field-ion microscope. (6)

experimental procedure because a variety of factors including pulse fraction, specimen temperature and background pressure in the system can markedly affect the quality and content of the data.

The original APFIM was a straight time-of-flight atom probe (2). A significant improvement in the mass resolution of the atom probe mass spectra was accomplished by incorporating a Poschenrieder energy-compensating lens to the time-of-flight mass spectrometer (3). Additional information concerning the theory of field-ion microscopy and atom probe analysis can be found in Refs. 4-6. The various types of atom probes are described in an excellent text on the technique by Miller & Smith (6). The instruments include the imaging atom probe of Panitz (7) and the pulsed laser atom probe (PLAP) developed by Kellogg and Tsong (8). The PLAP permits the analysis of semiconductors and other materials which have high electrical resistivities by using a pulsed laser beam as opposed to a high voltage pulse for field-evaporation of the surface atoms.

One of the recent developments in APFIM instrumentation over the past 7 years has been the incorporation of a reflectron energy-compensated mirror as a replacement for the Poschenrieder lens in the atom probe. Drachsel, et al. developed and tested a reflectron system for atom probe applications in 1988 (9). Further improvements were accomplished by Camus and Melmed with their development of a reflectron energy-compensating mirror with einzel lens for focussing the ions on the detector (10). They noted a significant improvement in mass resolution over the conventional straight time-of-flight atom probes. The reflectron system is the basis of a commercially produced atom probe, the VSW APFIM 200 designed by Waugh and co-workers (11).

Within the past 7 years, there have been significant advances in the performance of the APFIM with the introduction of the position sensitive atom probe (PoSAP) by Cerezo and co-workers at Oxford(12,13), and the tomographic atom probe (TAP) by Blavette and colleagues in Rouen (14-16). As opposed to the conventional selected area atom probe analysis which analyzes a narrow "cylinder" of material, the new instruments (POSAP and TAP) permit the 3-dimensional analysis of the sample by storing the position of the field-evaporated ion as well as its time-of-flight. The acquired data are stored in a computer so that further detailed analyses of morphology, size, and distribution of various features can be performed. This capability strengthens the APFIM technique by dramatically increasing the amount of data obtainable from a single needle specimen and by permitting the reconstruction of the analyzed specimen.

The PoSAP, developed in 1988 by Cerezo and co-workers at Oxford, involves the use of a wedge-and-strip position-sensitive detection system, shown schematically in Figure 2 (12). However, only one ion per pulse can be accurately detected. Therefore, slow field-evaporation rates are necessary for the acquisition of data. The instrument has been successfully applied to the analysis of Cu precipitation in Fe-Cu alloys (17), precipitation in Al alloys (18) and spinodal decomposition in Fe-Cr (19) and Cu-Co alloys (20). Unfortunately, the mass resolution of the PoSAP analysis is degraded over the conventional straight time-of-flight atom probes. Cerezo and colleagues are continuing

to improve the mass resolution and multiple ion detection ability of the PoSAP (21).

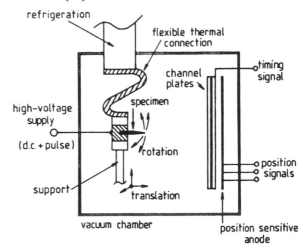

Figure 2. Schematic diagram of the PoSAP developed by Cerezo et al. (13)

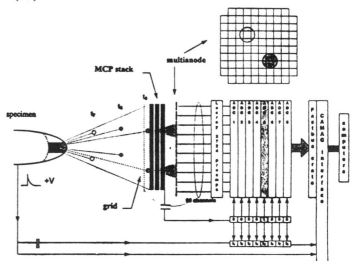

Figure 3. Schematic diagram of the TAP developed by Blavette et al. (15)

In 1993, Blavette and colleagues at the Université de Rouen developed the TAP using a 10 X 10 independent anode array as the detection system, therefore permitting the analysis of multiple ions on a single voltage pulse (15,16). Figure 3 shows the configuration of the TAP. This development greatly enhanced the rate of data collection in that field-evaporation rates similar to conventional AP operation may be used. A range of materials, including several Ni-base superalloys (22,23) and neutron-irradiated pressure vessel steels (24), have been examined using the TAP. Furthermore, striking examples illustrating the excellent depth resolution of the technique have been obtained. Figure 4, from the work of Deconihout, et al.(25), shows a graphical representation of the analysis along a [001] pole in a γ' precipitate in a Ni-base superalloy. The Al-rich and Ni-rich planes are clearly visible in this true "lattice image" of the precipitate.

Computer data acquisition systems permit the re-analysis of the 3D atom probe data in a variety of options, including compositional profiles through features in various directions, and the manipulation of the analyzed volume to "see" the features from other perspectives. Considerable effort has been expended in the calibration and analysis of 3D atom probe data (22-29).

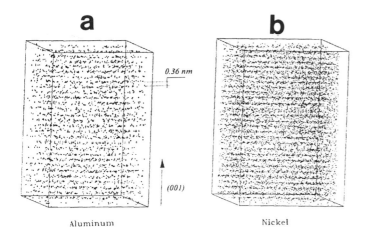

Figure 4. Two-dimensional representation of TAP data obtained from the analysis through a [001]-oriented γ' precipitate in a Ni-base superalloy by Deconihout et al.(25) (a) TAP data showing the arrangement of the Al ions, and (b) TAP data showing the alignment of the Ni ions within the γ' precipitate. Note that the depth spacing is one (001) plane.

To maximize analytical capabilities, Hono and co-workers have designed and built an APFIM which incorporates both the reflectron-type time-of-flight atom probe (TOFAP) with a PoSAP (30). The reflectron TOFAP permits conventional high resolution atom probe microanalysis whereas the PoSAP provides the 3D analysis of the material. This idea of a combined APFIM instrument will undoubtedly become more widespread.

The major limitations of the APFIM technique include the limited sampling volume for analysis, and the tendency of APFIM specimens to fracture during examination due to the high stresses associated with the high positive voltage applied to the specimen. APFIM analysis can be time-consuming as compared with other microanalytical technique. Also, features for analysis must generally be present in a number density $>\sim 10^{17}$ per cm^3. Locating specific microstructural features (such as grain boundaries) for analysis requires the pre-examination of the APFIM specimens by another technique such as TEM. In spite of these limitations, the data analysis from AP microanalysis is relatively straightforward, and there is no restriction on the elements which can be analyzed. Furthermore, the excellent spatial resolution of this technique permits the quantitative analysis of features as fine as 1 nm.

Although there are impressive technical achievements concerning 3D APFIM analysis (31), the conventional energy-compensated APFIM remains an important analytical tool for the analysis of ultra-fine scale features in materials. However, it is necessary to note the limitations of the various analytical techniques, and to incorporate a variety of appropriate techniques

in the microstructural characterization of materials. It is through microstructure that the mechanical and physical properties of a material are established. Understanding the relationship between a material's microstructure and properties, and understanding the development of a material's microstructure both during thermomechanical processing and during service (either under high temperature or neutron irradiation) allows one to tailor or develop improved materials for a particular application. The following examples represent some practical applications of AEM and APFIM in the characterization of the microstructure of several Ni- and Fe-base alloys.

Ni-Base Superalloys

Alloy X-750: AEM is used for precipitate identification and morphological analysis. APFIM provides microcompositional data for γ' and γ.

Alloy X-750 is a γ'-strengthened superalloy which is used in a variety of applications in the commercial nuclear industry because of its superior resistance to stress corrosion cracking (SCC)(32). This material is generally solution-annealed at ~1090°C followed by an aging treatment at ~720°C for 20 h. This aging treatment produces a relatively uniform distribution of $L1_2$-ordered γ' precipitates with a rounded cuboidal morphology throughout the matrix, and discrete Cr-rich $M_{23}C_6$ carbides along

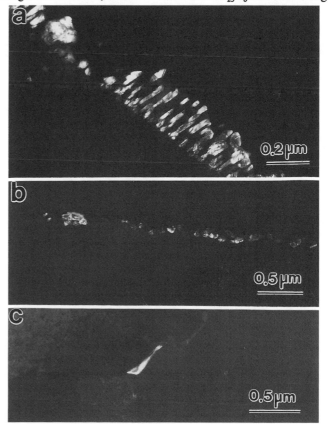

Figure 5. Dark-field transmission electron micrographs showing the typical intergranular precipitates observed in Alloy X-750 after aging at 704°C for 24 h. (a) "Cellular" $M_{23}C_6$; (b) discrete $M_{23}C_6$; and (c) M_7C_3.

the grain boundaries. Additionally, Cr-rich M_7C_3 carbides and, depending upon the boron content of the alloy, Ni and B-rich $M_{23}X_6$ precipitates can form at grain boundaries during the aging treatment. AEM permits the detailed characterization of intergranular precipitation in these alloys, the extent of which is frequently associated with susceptibility to SCC. The various precipitates formed in thermally-treated Alloy X-750 are shown in Figure 5. (33) Yonezawa, et al. have shown the importance of intergranular carbide morphology on the stress corrosion cracking behavior of the material in light water reactor environments (32).

The cooling rate (fast air-cool vs. water-quench) from the solution-annealing temperature has been shown to affect the microstructure developed during the subsequent aging treatment and, in turn, the material's resistance to SCC (33). To understand the details of the microstructural development, AEM and APFIM

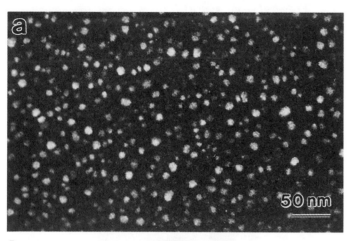

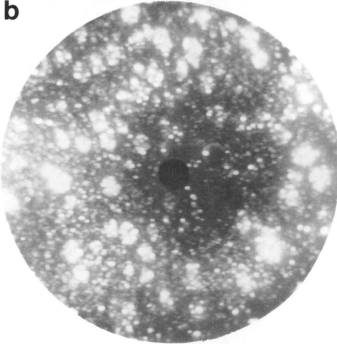

Figure 6. Complementary (a) dark-field transmission electron and (b) field-ion micrographs containing brightly-imaging γ' precipitates formed on fast air-cooling from the solution-annealing temperature.

Figure 7. Bright-field transmission electron micrograph of the as-water-quenched Alloy X-750 sample containing dislocations associated with the quenching stresses.

techniques were employed. Alloy X-750 specimens which had been solution-annealed for 1h at 1093°C were either fast air-cooled or water-quenched prior to aging at 718°C. These materials were examined in the as-cooled and as-aged conditions. Fast air-cooling promoted the intragranular precipitation of very fine (~ 5-10 nm) γ' and ~ 10-50 nm intergranular Cr-rich $M_{23}C_6$ carbides. Complementary dark-field transmission electron and field-ion micrographs of the very fine γ' precipitates are presented in Figure 6. The water-quenched specimens contained no γ' precipitates or intergranular carbides. However, dislocations associated with quenching stresses were evident in the TEM examination, Figure 7.

Subsequent aging of the as-solution-annealed specimens resulted in subtle differences in terms of the intergranular carbide precipitation. The fast air-cooled samples were characterized by the presence of discrete intergranular Cr-rich $M_{23}C_6$ carbides, groups of which were semicoherent with one of the two γ grains. In the water-quenched specimens, AEM examination revealed that there was notable "cellular" (discontinuous) carbide precipitation, as shown in Figure 8. Dark-field TEM and FIM micrographs showed no significant difference in the extent or size of the γ' precipitates between the samples. The fine size of these intragranular precipitates precludes their chemical analysis by STEM-EDS microanalysis in the AEM. However, the high spatial resolution of the APFIM is well suited to this analysis. Atom probe microanalysis of the γ' precipitates in the fast air-cooled + aged and water-quenched + aged samples were comparable (34). The results of the AP analyses showed that these precipitates contained approximately: 1 Cr - 2 Fe - 15 Ti - 7 Al - 2 Nb - 0.3 Si - 0.1 Mn - bal Ni (at.%).

The effect of cooling rate on the subsequent microstructural development was attributed to an increased driving force for grain boundary migration which promoted the development of the "cellular" $M_{23}C_6$ carbides, which was associated with the "poor" (SCC-susceptible) microstructure classification of Yonezawa et al. (32).

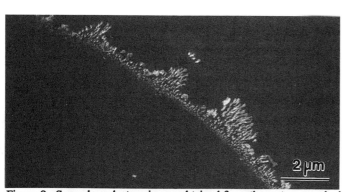

Figure 8. Secondary electron image obtained from the water-quenched + aged Alloy X-750 sample showing pronounced "cellular" or discontinuous precipitation of $M_{23}C_6$ carbides along grain boundaries.

Alloy 718: AEM provides chemical and structural information for second phase precipitates (>0.1 μm in size). APFIM is used for measurement of γ", γ' and γ compositions.

Alloy 718 is another Ni-base superalloy which possess good elevated temperature properties and, depending upon the thermal treatment, good resistance to SCC (35,36). The additions of Nb, Mo, Ti and Al in Alloy 718 promote the formation of two fine intragranular strengthening precipitates, γ" and γ'. The DO_{22}-ordered γ" precipitates are disc-shaped and are crystallographically related to the γ matrix: $(001)_{\gamma'} // \{100\}_\gamma$ and $[001]_{\gamma'} // <100>_\gamma$. The $L1_2$-ordered γ' precipitates are slightly rounded cuboids which exhibit the cube-cube orientation relationship with the γ matrix: $(001)_{\gamma'} // (001)_\gamma$ and $[100]_{\gamma'} // [100]_\gamma$. The strengthening precipitates are formed during multi-stage thermal treatments, originally developed for creep resistance in aerospace applications. The use of such thermal treatments, however, may produce microstructures which are susceptible to stress corrosion cracking (35). Furthermore, a wide variety of second phases can form in this alloy, the presence of which can adversely affect mechanical properties. These phases include DO_a-ordered δ (Ni_3Nb-type), Laves, and MC-type carbides. The microstructural development which occurs during the multistage treatments has been evaluated by AEM and APFIM (37). By characterizing the material after each stage in a typical multi-stage thermal treatment, it was possible to obtain information for clarifying the precipitation sequence in this alloy. With this information, appropriate heat treatments can be designed for optimization of the material's resistance to SCC.

AEM characterization of materials in the solution-annealed condition revealed the presence of various coarse blocky MC-type carbides and a second coarse blocky phase, most likely formed during solidification. The second phase was identified as C14 Laves phase with $a_o =~ 0.47$ nm and $c_o =~ 0.75$ nm by electron diffraction analysis. STEM-EDS microanalysis indicated that the composition of this phase was consistent with $(NiCr_{0.5}Fe_{0.5})_2(NbMo)$. Subsequent aging at 870°C resulted in the formation of coarse (~0.2 to 0.5 μm) γ" precipitates and in the precipitation of both inter- and intragranular δ needles (~0.5 to several μm in length). The composition of the δ precipitate was measured by STEM-EDS microanalysis as: ~19.5 Nb - 4 Ti - 2.3

Fe - 3.9 Cr - 70.3 Ni (at.%). AEM and APFIM microanalysis of the coarse γ" precipitates formed at 870°C yielded consistent compositions (at.%):
AEM: 21.3 Nb - 0.8 Al - 4.0 Ti - 3.3 Fe - 2.5 Cr - 0.4 Mo - bal Ni
APFIM: 23.8 Nb - 0.6 Al - 6.3 Ti - 2.2 Fe - 1.5 Cr - 0.7 Mo - bal Ni

The lower temperature age at 760°C promoted the precipitation of fine γ" and γ' (~10 to 40 nm in size) throughout the matrix, with slightly coarser γ" precipitates along high-angle grain boundaries. The fine size of the γ" and γ' precipitates precludes their compositional analysis by AEM. Although the γ" and γ' are generally assumed to be $Ni_3(NbTi)$ and $Ni_3(AlTi)$, respectively, APFIM microanalysis demonstrated that both precipitates are $Ni_3(NbTiAl)$, with varying proportions of Nb, Ti and Al, and also with no specific compositional limits between the Nb, Ti and Al.(37) The extent of γ" and γ' precipitation during the single 870°C and 760°C ages and during a double 870°C + 760°C age is shown in Figures 9-11. Figure 12 is a field-ion micrograph of the double-aged alloy containing both coarse and fine γ" and fine γ' precipitates.

Further AEM studies on Alloy 718 revealed that NbC(N) precipitation occurs during aging at 760°C. AEM analysis of both thin-foil specimens and carbon extraction replica specimens enabled the crystal structure and composition of the precipitates to be determined. (38) The thin intergranular carbide film, ~ 1μm in length, was nonuniformly distributed in the material, and was associated with the existence of narrow γ" and γ' precipitate-free zones.

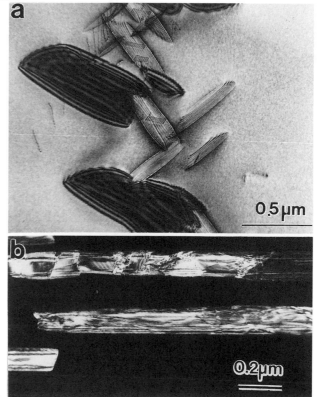

Figure 9. (a) Bright-field transmission electron micrograph of the coarse γ" precipitates, and (b) dark-field transmission electron micrograph of δ precipitates formed in Alloy 718 during aging at 870°C.

Although APFIM characterization is frequently associated with the analysis of fine-scale microstructural features such as precipitates, the ability of the technique to obtain quantitative matrix analyses is extremely useful in evaluating the precipitation potential during further aging. Knowledge of the solute content of the matrix is important for the prediction of the material's response to subsequent thermal treatment. With the microstructural data obtained by AEM and APFIM, the precipitation sequence in Alloy 718 was confirmed, and the ability to tailor the microstructure through thermomechanical processing was increased.

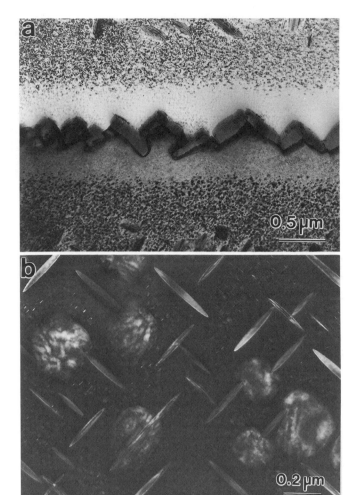

Figure 11. (a) Bright-field transmission electron micrograph showing intergranular δ precipitates and associated primary and secondary γ" precipitate-free zones in Alloy 718 after the 870+760°C double age. (b) Dark-field transmission electron micrograph showing all 3 <001> variants of the γ" precipitates formed in the double-aged specimen.

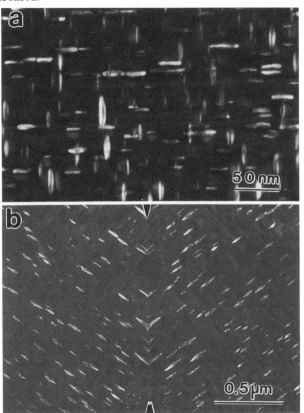

Figure 10. Dark-field transmission electron micrograph showing (a) 2 <001> variants of the γ" precipitates in a [001]-oriented foil, and (b) γ" precipitation at a coherent twin boundary in Alloy 718 aged at 760°C

Alloy 625: AEM provides general microstructural data and precipitate identification.

Alloy 625 is a γ"-strengthened superalloy which is attractive for applications where good mechanical properties and resistance to intergranular SCC and irradiation-assisted stress corrosion cracking (IASCC) are important (39). Like Alloys X-750 and 718, the mechanical properties and SCC resistance can be dramatically altered by heat treatments. This alloy can be precipitation-hardened by aging in the temperature range 550 - 800°C. In the direct-aged condition (wrought + ~660°C for 80 h), this material contains a relatively uniform distribution of fine (~ 15 nm in diameter) γ" precipitates throughout the grains and along numerous grain boundaries, Figure 13 (39). The dislocations associated with the thermomechanical processing are

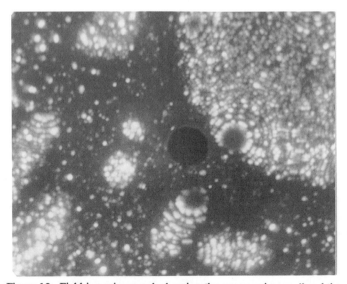

Figure 12. Field-ion micrograph showing the coarse primary γ" and the fine γ' and secondary γ" precipitates in Alloy 718 after aging at 870°C + 760°C.

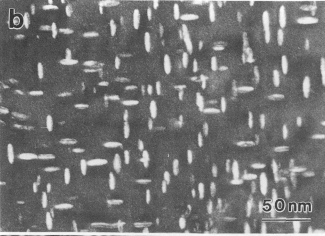

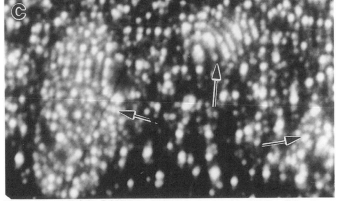

Figure 13. (a) Bright-field transmission electron micrograph of the direct-aged Alloy 625. (b) Dark-field transmission electron micrograph showing 2 of the 3 <001> variants of the intragranular γ" precipitates. (c) Field-ion micrograph of brightly-imaing γ" precipitates in a dark γ matrix.

present throughout the matrix of the direct-aged alloy, and act as nucleation sites for the γ" precipitates. The γ" precipitates exhibit the following orientation relationship with the matrix: $(001)_{\gamma"} // \{100\}_{\gamma}$ and $[001]_{\gamma"} // <100>_{\gamma}$ The field-ion image of Figure 13 shows the brightly imaging γ" precipitates in the dark matrix. Various second phases such as Cr-rich $M_{23}C_6$, Mo-enriched M_6C carbides, and a limited amount of δ (Ni_3Nb) are also formed during the thermal treatment. During neutron irradiation at 370°C, the γ" precipitates partially dissolve and a

new metastable phase is formed (39). Electron diffraction analysis was performed to evaluate the structure of this new phase. Figure 14 contains selected area electron diffraction patterns (SADPs) which clearly show the presence of additional reflections at the 1/3 and 2/3 (220) positions. The electron diffraction data indicate that this new metastable phase is consistent with the body-centered orthorhombic Pt_2Mo structure. This phase appears as fine (~ 5-10 nm) discrete precipitates distributed relatively uniformly throughout the matrix. Dark-field micrographs of the γ" and new metastable precipitates are shown in Figure 15. STEM-EDS microanalysis of the coarser agglomerated precipitates indicated slight enrichments in Mo and Nb which suggests that the new phase is $Ni_2(MoNb)$. APFIM characterization of these materials will provide additional microchemical data. Previous investigations of Ni-Mo binary alloys by Yamamoto, et al. (40), and Das and Thomas (41) have shown that a similar orthorhombic Pt_2Mo-type precipitate, Ni_2Mo, is a metastable phase which, along with Ni_4Mo, forms as disordered Ni_3Mo decomposes to the more stable ordered Ni_3Mo.

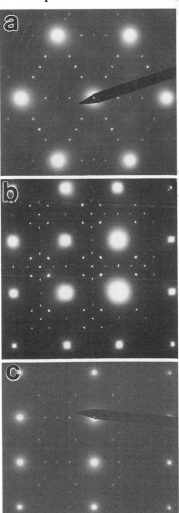

Figure 14. Selected area electron diffraction patterns obtained showing the presence of the new metastable Pt_2Mo-type phase in Alloy 625 after neutron irradiation at 370°C. (a) [111], (b) [001], and (c) [112] zone axes.

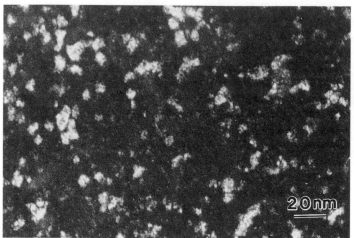

Figure 15. Dark-field transmission electron micrograph of the new metastable Pt₂Mo-type phase in Alloy 625.

Fe-Base Alloys

Pressure Vessel Steels: APFIM detects and analyzes ultra-fine solute clusters and precipitates, and intergranular segregation.

The APFIM technique has provided unique information concerning the microstructure of neutron irradiated pressure vessel steels and weldments as it relates to the degradation in material performance. An extensive amount of research has been performed to elucidate the mechanism of irradiation embrittlement of these steels, initially by vast compositional matrices to identify those elements which were associated with the degradation in impact properties of these materials. Through this work it is now known how to minimize pressure vessel steel embrittlement. Conventional AEM provided microstructural information on the type and extent of carbide precipitation, the identification of ε Cu precipitates, and inclusion analyses. APFIM was successfully applied to the analysis of Cu precipitates and Mo_2C in irradiated A302B steels by Miller and Brenner (42). The evaluation of neutron-irradiated A533B weld metal demonstrated for the first time the presence of solute-rich clusters containing Cu, Mn, and Ni (43). Figure 16 contains a field-ion micrograph of an irradiated A533B weldment and an evaporation sequence diagram obtained from the analysis through an irradiation-induced cluster in the weld. These clusters act as local hardening features and obstacles to dislocation movement. Further investigations into irradiation-induced microstructural changes in pressure vessel steels have demonstrated the unique capabilities of the APFIM to the evaluation of intergranular segregation, solute clustering, and the identification of ultra-fine (<3 nm) carbides and nitrides (44). Recently, Pareige, et al. have successfully used the TAP to characterize a pressure vessel steel after high fluence neutron irradiation (24). The TAP provided the first direct three-dimensional evidence of the solute-rich "clusters" in the irradiated alloy, thus confirming the conventional APFIM cluster observation. Additionally, they noted significant incorporation of Si into the clusters after high fluence (~10^{20} neutrons/cm², E> 1 MeV). Advances in data analysis, such as the mean separation

technique of Hetherington and Miller, provided the ability to quantify the significance of "solute clustering" in materials (45). In addition, the APFIM analysis of the irradiated matrix material reveals the solubility of Cu in the alloy. It is interesting to note that in a wide variety of Cu-bearing RPV steels and weldments, the matrix Cu content after neutron irradiation is remarkably constant (44). The data provided by APFIM concerning clustering and precipitation phenomena associated with neutron irradiation has been extended using field emission gun scanning transmission electron microscopy (FEG-STEM) EDS microanalysis. The high spatial resolution of the FEG-STEM technique has been successfully utilized in the microstructural characterization of irradiated pressure vessel steels by researchers at AEA Technology and Nuclear Electric (46).

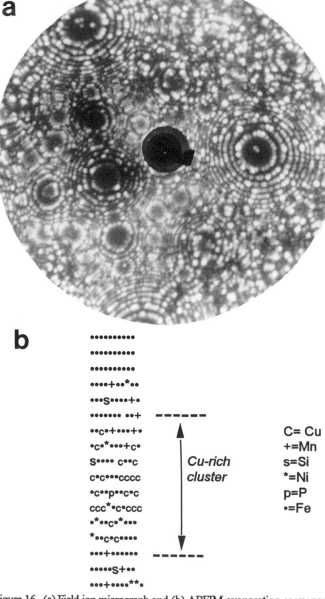

Figure 16. (a) Field-ion micrograph and (b) APFIM evaporation sequence diagram showing a Cu-rich solute cluster obtained from neutron-irradiated A533B-type weld metal.

98

Fe-Be Alloys: AEM provides crystallographic data and phase distributions. APFIM is used to measure the composition of phases and obtain 3-dimensional morphological information of the precipitates.

Dramatic hardening occurs in solution-annealed Fe - (15 to 25) at% Be alloys during aging within the temperature range 265 to ~400°C. The associated microstructural development which occurs during aging Fe-25at%Be alloys within the miscibility gap has been studied in detail using AEM and APFIM (47-49). These alloys decompose spinodally into a B2-ordered FeBe phase and a Be-depleted body-centered cubic (bcc) α Fe. Although the contrast between the two phases is very complex in bright-field images due to strain effects, the two phases can be clearly imaged using the dark-field technique. Figure 17 shows a dark-field image formed using a reflection from the Be-rich B2-ordered phase. In this image, the B2-ordered phase appears as discrete brightly-imaging cuboids in a dark α Fe matrix. Field-ion images of this material clearly shows the two phases due to differences in their field-evaporation behavior. In the FIM image of Figure 17, the contrast is reversed to that in the dark-field TEM micrograph; the B2-ordered phase is darkly-imaging whereas the Fe-rich phase is brightly-imaging. The 3-dimensional nature of the spinodally decomposed microstructure is clearly evident during the field-evaporation of the aged alloys which demonstrates that the interconnected phase is the B2-ordered Be-rich phase (47). Furthermore, the FIM data confirms that the discrete cuboidal (non-continuous) phase is in fact bcc α Fe, in contrast to the TEM images. This example illustrates the importance of using complementary techniques in microstructural analysis.

The two techniques have been equally important in the identification of two metastable phases in the Fe-Be system (48, 49). The crystallographic structures of the two unknown phases were obtained by both conventional and microdiffraction techniques. Electron diffraction patterns obtained from the unknown phases showed that the structures were crystallographically related to the disordered bcc α phase. The complementary compositional data were provided by atom probe microanalysis, which is well-suited to the measurement of Be in Fe. Importantly, APFIM microanalysis is the only analytical technique which can be used to obtain quantitative compositional data for these materials. The data obtained by both techniques were essential in the identification of the B32-ordered FeBe phase and the B8-ordered BeFe phase, Figure 18.

Summary

Microstructural characterization is an important aspect of materials research activities to support the development of improved material properties. Techniques such as AEM and APFIM can be most useful for the identification of phases, both structural and compositional, as well as the measurement of second phases, segregation, and the analysis of defects. Selection of the appropriate analytical technique by experienced microstructural analysts will depend on the particular problem to be addressed and the type of data required. The scale of analysis

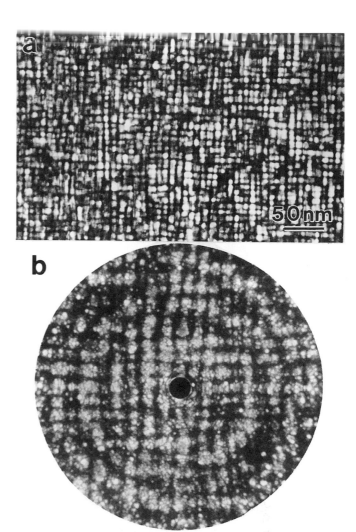

Figure 17. (a) Dark-field transmission electron micrograph formed using a B2 superlattice reflection, and (b) complementary field-ion micrograph of Fe-25at.%Be alloy aged at 400°C for 20 min. Note that the brightly-imaging phases in (a) and (b) are the Be-enriched B2-ordered phase. The FIM image (b) shows that the Be-enriched phase is continuous, with discrete Fe-rich precipitates.

is determined by the material and the particular properties to be evaluated. AEM has distinct advantages for the microstructural characterization of materials, providing crystallographic analysis of phases, microchemical analysis of features >50 nm in size, information on defect structures, and data on the size and extent of precipitation in materials. The advantages of APFIM include the quantitative microchemical analysis of ultra-fine (~1 nm) features, measurement of interfacial segregation, and morphological information on precipitates and other phases in electrically conductive materials. The benefits of both techniques have been illustrated through their combined use in the evaluation of Ni- and Fe-base alloys. It is necessary to note that all analytical techniques have their specific advantages and disadvantages; therefore, judicious selection of the appropriate technique or combination thereof is necessary for timely and thorough microstructural analysis. With the new developments in the APFIM instrumentation, there will be an increase in interest and materials applications as these instruments become more

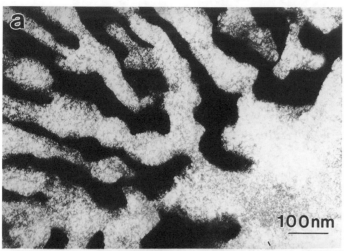

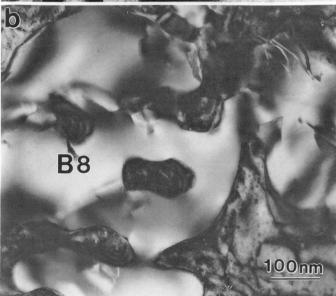

Figure 18. (a) Dark-field transmission electron micrograph of the B32-ordered FeBe phase, and (b) bright-field transmission electron micrograph of the B8-ordered BeFe phase formed in Fe-25at.%Be.

widespread. Thus, the importance of the combined technique approach to materials characterization will continue to grow.

Acknowledgments

I wish to thank Raman Jayaram of Oak Ridge National Laboratory (ORNL) for valuable discussions, P. Pareige, D. Blavette, A. Menand and P. Bas of the Université de Rouen, and A. Cerezo and G.D.W. Smith of Oxford University for their kind permission to use their extraordinary examples of 3D atom probe microanalysis. Also, much APFIM analysis has been performed over the past decade through my collaboration with M.K. Miller at ORNL, with whom I have shared many lively experiments and discussions over the years. I also wish to thank D.M. Symons and B.F. Kammenzind of the Bettis Atomic Power Laboratory for helpful comments and discussions.

References

1. D.B. Williams, **Practical Analytical Electron Microscopy in Materials Science,**Verlag Chemie International, Philips Electronic Instruments Inc. Electron Optics Publishing Group, (1984).

2. E.W. Müller, J.A. Panitz and S.B. McLane, **Rev. Sci. Instruments, 39,** 83 (1968).

3. W.P. Poschenrieder, **Int. J. Mass Spectrom. Ion Phys., 6,** 413 (1971).

4. E.W. Müller and T.T. Tsong. **Field-Ion Microscopy: Principles and Applications,** Elsevier, Amsterdam (1969).

5. K.M. Bowkett and D.A. Smith, **Field-Ion Microscopy,** North Holland, Amsterdam (1970).

6. M.K. Miller and G.D.W. Smith, **Atom Probe Microanalysis: Principles and Applications to Materials Problems,** Materials Research Society, Pittsburgh (1989).

7. J.A. Panitz, U.S. Patent No. 3,868,507, Feb. 25, 1975.

8. G.L. Kellogg and T.T. Tsong, **J. Applied Physics, 51,** 1184 (1980).

9. W. Drachsel, L. von Alvensleben and A.J. Melmed, **J. de Physique, 50,** C8-541 (1989).

10. P.P. Camus and A.J. Melmed, **Surface Sci., 246,** 415 (1991).

11. A.R. Waugh, C.H. Richardson and R. Jenkins, **Surface Sci., 266,** 501 (1992).

12. A. Cerezo, T.J. Godfrey and G.D.W. Smith, **Rev. Sci. Instruments, 59,** 862 (1988).

13. A. Cerezo, T. Godfrey and G.D.W. Smith, **J. de Physique, 49,** C6-25 (1988).

14. A. Bostel, D. Blavette, A. Menand and J.M. Sarrau, **J. de Physique, 50,** C8-501 (1989).

15. D. Blavette, B. Deconihout, A. Bostel, J.M. Sarrau, M. Bouet and A. Menand, **Rev. Sci. Instruments, 64 (10),** 2911 (1993).

16. D. Blavette, B. Deconihout, A. Bostel, J.M. Sarrau and A. Menand, **Nature, 363,** 432 (1993).

17. A. Cerezo, T.J. Godfrey, C.R.M. Grovenor, M.G. Hetherington, R.M. Hoyle, J.P. Jakubovics, J.A. Liddle, G.D.W. Smith and G.M. Worral, **J. Microscopy, 154,** 215 (1989).

18. P.J. Warren, C.R.M. Grovenor and J.S. Crompton, **Surface Sci., 266,** 342 (1992).

19. A. Cerezo, M.G. Hetherington, J.M. Hyde, and M.K. Miller, **Scripta Met., 25,** 1435 (1991).

20. R.P. Setna, J.M. Hyde, A. Cerezo, G.D.W. Smith and M.F. Chisholm, **Applied Surface Sci., 67,** 368 (1993).

21. A. Cerezo, T.J. Godfrey, J.M. Hyde, S.J. Sijbrandij and G.D.W. Smith, **Applied Surface Sci.**, **76/77**, 374 (1994).

22. B. Deconihout, A. Bostel, A. Menand, J.M. Sarrau, M. Bouet, S. Chambreland and D. Blavette, **Applied Surface Sci.**, **67**, 444 (1993).

23. B. Deconihout, A. Bostel, P. Bas, S. Chambreland, L. Letellier, F. Danoix and D. Blavette, **Applied Surface Sci.**, **76/77**, 145 (1994).

24. P. Pareige, J.C. Van Duysen and P. Auger, **Applied Surface Sci.**, **67**, 342 (1993).

25. B. Deconihout, A. Bostel, M. Bouet, J.M. Sarrau, P. Bas and D. Blavette, **Applied Surface Sci.**, **87/88**, 428 (1995).

26. M.G. Hetherington, A. Cerezo, J.M. Hyde and G.D.W. Smith, **Surface Sci.**, **266**, 463 (1992).

27. J.M. Hyde, A. Cerezo, R.P. Setna, P.J. Warren and G.D.W. Smith, **Applied Surface Sci.**, **76/77**, 382 (1994).

28. P.P. Camus, D.J. Larson and T.F. Kelly, **Applied Surface Sci .**, **87/88** , 305 (1995).

29. P. Bas, A. Bostel, B. Deconihout and D. Blavette, **Applied Surface Sci.**, **87/88**, 298 (1995).

30. K. Hono, R. Okano, T. Saeda and T. Sakurai, **Applied Surface Sci.**, **87/88**, 453 (1995).

31. **Applied Surface Sci.**, **76/77**, xxi-xxvi (1994).

32. T. Yonezawa, K. Onimura, N. Sakamoto, N. Sasaguri, H. Nakata and H. Susukida, **1st International Symposium on Environmental Degradation of Materials in Nuclear Power Systems-- Water Reactors**, 345 (NACE, 1984).

33. M.G. Burke, I.L.W. Wilson and T.R. Mager, **5th International Symposium on Environmental Degradation of Materials in Nuclear Power Systems-- Water Reactors**, ed. E.P. Simonen 287(ANS, 1992).

34. M.K. Miller and M.G. Burke, **Applied Surface Sci.**, **67**, 292 (1993).

35. I.L.W. Wilson and M.G. Burke, **Superalloys 718, 625 and Various Derivatives**, ed. E.A. Loria, 681 (TMS, 1991).

36. M.G. Burke, T.R. Mager, M.T. Miglin, and J.L. Nelson, **Superalloys 718, 625, 706 and Various Derivatives**, ed. E.A. Loria, 763 (TMS, 1994).

37. M.G. Burke and M.K. Miller, **Superalloys 718, 625 and Various Derivatives**, ed. E.A. Loria, 337 (TMS, 1991).

38. M.G. Burke, T.R. Mager and J.L. Nelson, **6th International Symposium on Environmental Degradation of Materials in Nuclear Power Systems-- Water Reactors**, eds. R. E. Gold and E.P. Simonen, 821 (TMS, 1993).

39. R. Bajaj, W.J. Mills, M. Lebo, B.Z. Hyatt and M.G. Burke, **7th International Symposium on Environmental Degradation of Materials in Nuclear Power Systems-- Water Reactors**, eds. G.P. Airey et al., 1093 (NACE, 1995).

40. M. Yamamoto, S. Nenno, T. Saburi and Y. Mizutani, **Trans. J.I.M.**, **11**, 120 (1970).

41. S. Das and G. Thomas, **Phys. Stat. Sol. (a) 21**, 177 (1974).

42. M.K. Miller and S.S. Brenner, **Res Mechanica, 10,** 161 (1984).

43. M.G. Burke and S.S. Brenner, **J. de Physique, 47**, C2-239 (1986).

44. M.K. Miller and M.G. Burke, **J. Nuclear Materials,195,** 68 (1992).

45. M.G. Hetherington and M.K. Miller, **J. de Physique, 48**, C6-559 (1987).

46. C.A. English, W.J. Phythian, J.T. Buswell, J.R. Hawthorne and P.H.N. Ray, **15th International Symposium on Effects of Radiation on Materials - STP 1125**, eds. R.E. Stoller, A.S. Kumar and D.S. Gelles, 93 (ASTM, 1992).

47. M.G. Burke and M.K. Miller, **Proc. 43rd Annual Meeting of EMSA,** ed. G.W. Bailey, 70 (San Francisco Press, 1985).

48. M.K. Miller, M.G. Burke, S.S. Brenner, W.A. Soffa, K.B. Alexander and D.E. Laughlin, **Scripta Met., 18,** 285 (1984).

49. M.G. Burke and M.K. Miller, **Ultramicroscopy, 30**, 199 (1989).

101

Current Applications of X-ray Diffraction Residual Stress Measurement

Paul S. Prevey
Lambda Research, Cincinnati, OH

Abstract

A brief theoretical development of x-ray diffraction residual stress measurement is presented emphasizing practical engineering applications of the plane-stress model, which requires no external standard. Determination of the full stress tensor is briefly described, and alternate mechanical, magnetic, and ultrasonic methods of residual stress measurement are compared.

Sources of error arising in practical application are described. Subsurface measurement is shown to be necessary to accurately determine the stress distributions produced by surface finishing such as machining, grinding, and shot peening, including corrections for penetration of the x-ray beam and layer removal.

Current applications of line broadening for the prediction of material property gradients such as yield strength in machined and shot peened surfaces, and hardness in steels are presented. The development of models for the prediction of thermal, cyclic, and overload residual stress relaxation are described.

X-RAY DIFFRACTION (XRD) STRESS MEASUREMENT can be a powerful tool for failure analysis or process development studies. Quantifying the residual stresses present in a component, which may either accelerate or arrest fatigue or stress corrosion cracking, is frequently crucial to understanding the cause of failure. Successful machining, grinding, shot peening, or heat treatment may hinge upon achieving not only the appropriate surface finish, dimensions, case depth or hardness, but also a residual stress distribution producing the longest component life. The engineer engaged in such studies can benefit by an understanding of the limitations and applications of XRD stress measurement. This paper presents a brief development of the theory and sources of error, and describes recent applications of material property prediction and residual stress relaxation.

Application of XRD stress measurement to practical engineering problems began in the early 1950's. The advent of x-ray diffractometers and the development of the plane-stress residual stress model allowed successful application to hardened steels (1,2). The development of commercial diffractometers and the work of the Fatigue Design and Evaluation Committee of the SAE (3) resulted in widespread application in the automotive and bearing industries in the 1960's. By the late 1970's XRD residual stress measurement was routinely applied in aerospace and nuclear applications involving fatigue and stress corrosion cracking of nickel and titanium alloys, as well as aluminum and steels. Today, measurements are routinely performed in ceramic, intermetallic, composite, and virtually any fine grained crystalline material. A variety of position sensitive detector instruments allow measurement in the field and on massive structures. The theoretical basis has been expanded to allow determination of the full stress tensor, with certain limitations.

Stress is an extrinsic property, and must be calculated from a directly measurable property such as strain, or force and area. The available methods of residual stress "measurement" may be classified into two groups: those that calculate stress from strain assuming linear elasticity, and those that monitor other nonlinear properties.

In x-ray and neutron diffraction methods, the strain is measured in the crystal lattice, and the residual stress producing the strain is calculated, assuming a linear elastic distortion of the crystal lattice. The mechanical linear-elastic methods (dissection techniques) monitor changes in strain caused by sectioning, and are limited by simplifying assumptions concerning the nature of the residual stress field and sample geometry. Center hole drilling is more widely applicable, but is limited to stresses less than nominally 60% of the yield strength (4). All mechanical methods are necessarily destructive, and cannot be directly checked by repeat measurement.

All non-linear-elastic methods, such as ultrasonic and Barkhausen noise are subject to error from preferred orientation, cold work, and grain size. All require stress-free reference samples, which are otherwise identical to the sample under investigation, and are generally not suitable for laboratory residual stress determination at their current state of development.

XRD residual stress measurement is applicable to fine grained crystalline materials that produce a diffraction peak of suitable intensity, and free of interference in the high back-reflection region for any orientation of the sample surface. Surface measurements are nondestructive. Both the macroscopic residual stresses and line broadening caused by microstresses and damage to the crystals can be determined independently.

Macroscopic stresses, or macrostresses, which extend over distances large relative to the grain size of the material, are the stresses of general interest in design and failure analysis. Macrostresses are tensor quantities, and are determined for a given location and direction by measuring the strain in that direction at a single point. Macrostresses produce uniform distortion of many crystals simultaneously, shifting the angular position of the diffraction peak selected for residual stress measurement.

Microscopic stresses, or microstresses, are treated as scaler properties of the material, related to the degree of cold working or hardness, and result from imperfections in the crystal lattice. Microstresses arise from variations in strain between the "crystallites" bounded by dislocation tangles within the grains, acting over distances less than the dimensions of the crystals. Microstresses vary from point to point within the crystals, producing a range of lattice spacing and broadening of the diffraction peak.

Because the elastic strain changes the mean lattice spacing, only elastic strains are measured by x-ray diffraction. When the elastic limit is exceeded, further strain results in dislocation motion, disruption of the crystal lattice, and an increase in line broadening. Although residual stresses are caused by nonuniform plastic deformation, all residual macrostresses remaining after deformation are necessarily elastic.

The residual stress determined using x-ray diffraction is the arithmetic average stress in a volume of material defined by the irradiated area, which may vary from square centimeters to less than a square millimeter, and the depth of penetration of the x-ray beam. The linear absorption coefficient of the material for the radiation used governs the depth of penetration. For the techniques commonly used for iron, nickel, and aluminum alloys, 50% of the radiation is diffracted from a layer less than 5 μm deep. The shallow depth of penetration and small irradiated area allows measurement of residual stress distributions with spatial and depth resolution exceeding all other methods.

THEORY

A thorough development of the theory of x-ray diffraction residual stress measurement is beyond the scope of this paper. The interested reader is referred to the textbooks and general references (3,5,11,14). As in all diffraction methods, the lattice spacing is calculated from the diffraction angle, 2θ, and the known x-ray wavelength using Bragg's Law. The precision necessary for strain measurement in engineering materials can be achieved using diffraction peaks produced in the high back reflection region, where $2\theta > 120$ deg. The macrostrain is determined from shifts typically less than one degree in the mean position of the diffraction peak. The microstresses and crystallite size reduction caused by plastic deformation is usually expressed simply in terms of diffraction peak angular width, which may range from less than 0.5 deg. for annealed material to over 10 deg. in a hardened steel.

Plane-Stress Elastic Model. Because the x-ray penetration is extremely shallow (< 10 μm), a condition of plane-stress is assumed to exist in the diffracting surface layer. The stress distribution is then described by principal stresses σ_{11}, and σ_{22} in the plane of the surface, with no stress acting perpendicular to the free surface, shown in Figure 1. The normal component σ_{33} and the shear stresses $\sigma_{13} = \sigma_{31}$ and $\sigma_{23} = \sigma_{32}$ acting out of the plane of the sample surface are zero. A strain component perpendicular to the surface, ϵ_{33}, exists as a result of the Poisson's ratio contractions caused by the two principal stresses.

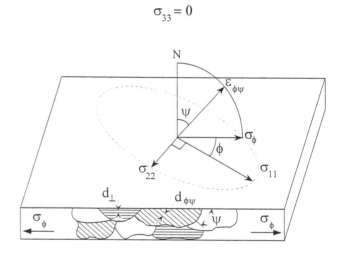

Fig. 1 - Plane stress at a free surface showing the change in lattice spacing with tilt ψ for a uniaxial stress σ_ϕ parallel to one edge.

104

The strain in the sample surface at an angle ϕ from the principal stress σ_{11} is then given by:

$$\epsilon_{\phi\psi} = \left(\frac{1 + \nu}{E}\right) \sigma_\phi \sin^2\psi - \left(\frac{\nu}{E}\right)(\sigma_{11} + \sigma_{22})$$ (Eq 1)

Equation 1 relates the surface stress σ_ϕ, in any direction defined by the angle ϕ, to the strain, $\epsilon_{\phi\psi}$, in the direction (ϕ,ψ) and the principal stresses in the surface. If $d_{\phi\psi}$ is the spacing between the lattice planes measured in the direction defined by ϕ and ψ, the strain can be expressed in terms of changes in the spacing of the crystal lattice:

$$\epsilon_{\phi\psi} = \frac{\Delta d}{d_0} = \frac{d_{\phi\psi} - d_0}{d_0}$$ (Eq 2)

where d_0 is the stress-free lattice spacing. Substituting into Eq. 1 and solving for $d_{\phi\psi}$ yields:

$$d_{\phi\psi} = \left[\left(\frac{1 + \nu}{E}\right)_{(hkl)} \sigma_\phi d_0\right] \sin^2\psi$$
$$- \left(\frac{\nu}{E}\right)_{(hkl)} d_0 (\sigma_{11} + \sigma_{22}) + d_0$$ (Eq 3)

where the appropriate elastic constants $(1 + \nu)/E_{(hkl)}$ and $(\nu/E)_{(hkl)}$ are now in the crystallographic direction normal to the (hkl) lattice planes in which the strain is measured. Because of elastic anisotropy, the elastic constants in the (hkl) direction commonly vary as much as 40% from the published mechanical values (5,6).

Equation 3 is the fundamental relationship between lattice spacing and the biaxial stresses in the surface of the sample. The lattice spacing $d_{\phi\psi}$, is a linear function of $\sin^2\psi$. Figure 2 shows the variation of d(311) with $\sin^2\psi$, for ψ ranging from 0 to 45° for shot peened 5056-O aluminum having a surface stress of -148 MPa (-21.5 ksi).

The intercept of the plot at $\sin^2\psi = 0$ equals the unstressed lattice spacing, d_0, minus the Poisson's ratio contraction caused by the sum of the principal stresses:

$$d_{\phi 0} = d_0 - \left(\frac{\nu}{E}\right)_{(hkl)} d_0(\sigma_{11} + \sigma_{22})$$
$$= d_0\left[1 - \left(\frac{\nu}{E}\right)_{(hkl)} (\sigma_{11} + \sigma_{22})\right]$$ (Eq 4)

The slope of the plot is:

$$\frac{\partial d_{\phi\psi}}{\partial \sin^2\psi} = \left(\frac{1 + \nu}{E}\right)_{(hkl)} \sigma_\phi d_0$$ (Eq 5)

which can be solved for the stress σ_ϕ:

$$\sigma_\phi = \left(\frac{E}{1 + \nu}\right)_{(hkl)} \frac{1}{d_0} \left(\frac{\partial d_{\phi\psi}}{\partial \sin^2\psi}\right)$$ (Eq 6)

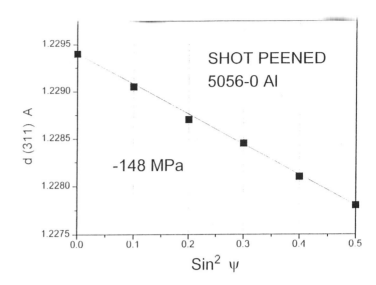

Fig. 2 - Linear dependence of d (311) upon $\mathrm{Sin}^2\psi$ for shot peened 5056-0 aluminum. Ref. (14)

The x-ray elastic constants can be determined empirically (6), but the unstressed lattice spacing, d_0, is generally unknown. However, because $E \gg (\sigma_{11} + \sigma_{22})$, the value of $d_{\phi 0}$ from Eq. 4 differs from d_0 by not more than $\pm 0.1\%$, and σ_ϕ may be approximated to this accuracy by substituting $d_{\phi 0}$ for d_0 in Eq. 6. The method then becomes a differential technique, and *no stress-free reference standards are required to determine d_0 for the plane-stress model*. All of the common variations of x-ray diffraction residual stress measurement, the "single-angle", "two-angle", and "$\sin^2\psi$" techniques, assume plane-stress at the sample surface, and are based on the fundamental relationship between lattice spacing and stress given in Eq. 3.

Stress Tensor Determination. An expression for the lattice spacing can be formulated as a function of ϕ and ψ, for the general case, assuming stresses exist normal to the surface (7). Nonlinearities producing separation of the $+\psi$ and $-\psi$ data in the form of elliptical curvature of the d-$\sin^2\psi$ plots termed "ψ splitting" are occasionally observed at the surface of ground hardened steels, and are attributable to shear stresses acting normal to the surface (8). Determination of the full stress tensor has been the focus of most academic research into XRD stress measurement over the last decade, and is necessary in all neutron diffraction applications because of the deep penetration into the sample.

In principle, the full stress tensor can be determined (7, 8). However, unlike the plane-stress model, the stress-free lattice spacing, d_o, must be known independently to the accuracy required for strain measurement (1 part in 10^5) in order to calculate the three normal stress components, σ_{11}, σ_{22}, and σ_{33}. Errors in the normal stress components, which are of primary interest, are proportional to the difference between the value of d_0 assumed and $d_{\phi 0}$. Large errors in both magnitude and sign of the three normal stress components can easily arise from errors in d_0. In most practical applications, such as the surfaces generated by machining, grinding, or hardening, the lattice spacing varies as a result of plastic deformation or heat treatment, precluding independent determination of the unstressed lattice spacing with sufficient precision (9-11). Further, other sources of nonlinearities in d-$\sin^2\psi$ plots such as subsurface stress gradients, instrument misalignment, and failure of the diffraction peak location method must first be eliminated (12,13). The full stress tensor method is therefore limited primarily to research applications.

SOURCES OF ERROR

Because XRD residual stress determination requires precision in the measurement of the angular position of diffraction peak on the order of 1 part in 10^5, many sources of error must be controlled. A thorough discussion of error is beyond the scope of this paper and have been addressed (3,5,11). The sources of error of primary importance in engineering applications may be placed in three categories: sample dependent errors, analytical errors, and instrumental errors.

Sample dependent errors may arise from an excessively coarse grain size, severe texture, or interference of the sample geometry with the x-ray beam. Both surface or subsurface stress gradients are common in machining and grinding, and may cause errors as high as 500 MPa, even altering the sign of the surface stress. Corrections can be made for penetration of the x-ray beam into the subsurface stress gradient using electropolishing to remove layers in fine increments on the order of 5-10 μm (3,1).

Electropolishing for subsurface measurement will cause stress relaxation in the layers exposed. If the stresses in the layers removed are high and the rigidity of the sample is low, the relaxation can be on the order of hundreds of MPa. For simple geometries and stress fields, closed form solutions are available (15). Recently, finite element corrections have been applied to arbitrary geometries (16).

Analytical errors may arise from the validity of the stress model assumed, the use of inaccurate elastic constants, or the method of diffraction peak location. Diffraction peaks several degrees wide must be precisely located within 0.01 deg. Various methods have been developed, but the fitting of Pearson VII functions to separate the Kα doublet and allow for peak defocusing caused by the change in ψ angle and line broadening as layers are removed in subsurface measurement is superior (12,13). X-ray elastic constants may be determined empirically to ASTM E1426 to an accuracy on the order of $\pm 1\%$ in four-point bending (6).

Instrumental errors arise from the misalignment of the diffraction apparatus or displacement of the specimen. Sample displacement from the center of the goniometer is the primary instrumental error. Divergence of the x-ray beam and sample displacement can cause "ψ splitting" which is indistinguishable in practice from the presence of sheer stresses, σ_{13} and σ_{23}, acting out of the surface. ASTM E915 provides a simple procedure using a zero stress powder to verify the instrument alignment, except for the accuracy of the ψ rotation.

SUBSURFACE MEASUREMENT

X-ray diffraction residual stress measurements are nondestructive. Attempts have been made since the 1960's to use nondestructive XRD residual stress measurement for process control, with limited success. The difficulty arises for two fundamental reasons.

First, surface residual stresses simply are not reliably representative of either the processes by which they were produced, or the stresses below the surface (17). Grinding and shot peening will commonly produce nearly identical levels of surface compression. Complete ranges of shot peening intensities will often produce virtually identical surface stresses with large differences in the depth of the compressive layer. Many surface finishing processes like tumbling, wire brushing, sand blasting, etc. will produce nearly identical surface compression which may mask subsurface tensile stresses due to welding, prior grinding, etc. Further, nondestructive surface measurements can not be corrected for potentially large errors due to penetration of the x-ray beam into a stress gradient.

Second, extensive studies have demonstrated that the subsurface peak residual stress rather than the surface residual stress generally governs fatigue life (18). The surface residual stress produced by turning, milling and grinding of steels, nickel, titanium, and aluminum alloys has been found to be the most variable and least characteristic of the machining process. The subsurface peak stress, either tensile or compressive, correlates with both room and elevated temperature fatigue behavior in extensive studies of surface integrity. The subsurface residual stress distribution must generally be obtained to adequately characterize a manufacturing process.

PROPERTY PREDICTION
FROM LINE BROADENING

The breadth of the diffraction peak used for residual stress measurement increases as materials are cold worked, or as a result of phase transformations such as hardening of martensitic steels. The broadening is primarily the result of two related phenomena: a reduction of the "crystallite" or coherent diffracting domain size, and nn increase in the range of microstrain. As a material is cold worked, or strained as a result of phase transformations, the perfect crystalline regions between dislocation tangles become smaller. When these regions are reduced to less than nominally $0.1\mu m$, the diffraction peak breadth increases with further reduction. The microstrain in each crystallite will vary about the mean value for the aggregate of such regions making up the polycrystalline body. This range of microstrain results in variation in lattice spacing of the diffracting crystallites, and increased line broadening. Other imperfections such as stacking faults and point defects also contribute to the peak breadth.

The relative contributions of crystallite size and microstrain to the integral breadth can be separated by the Warren-Averbach method (19). However, the separation is of little practical use in engineering applications, requires extensive data collection, and is subject to variations in interpretation. The measured peak breadth, even without correction for instrumental broadening, can be related directly to material properties of practical engineering interest such as the alteration of yield strength in cold worked alloys, and hardness in martensitic steels.

The hardness in martensitic steels can be measured simultaneously with residual stress with depth resolution on the order of the 5 μm penetration depth of the x-ray beam. The high depth resolution allows detection of thin work softened layers produced by deformation at the surface of critical components such as gears and bearings. An empirical relationship between the (211) peak breadth and

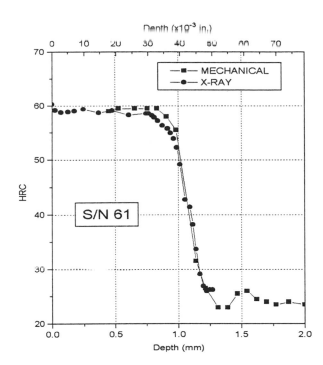

Fig. 4 - Comparison of mechanical (Vickens 500g) and XRD hardness measured at adjacent locations on an induction hardened SAE 1552 steel gear tooth. Ref. (16)

Rockwell C hardness for SAE 1552 steel is shown in Figure 3 (16). The hardness calculated from peak broadening is compared to mechanical microhardness measurements at an adjacent location on an induction hardened gear tooth in Figure 4. The high resolution of the x-ray diffraction technique allows a clear definition of the hardness gradient through the case-core interface.

The degree to which materials have been cold worked can be estimated from the peak breadth. If "cold work" is defined as the true plastic strain, a true stress-strain curve can then be used to estimate the resulting change in yield strength (20,21). An example of the relationship between the (420) diffraction peak width and the percent cold work (true plastic strain) for the nickel-base super alloy Rene 95 is shown in Figure 5. The results indicate the accumulated peak breadth is independent of the mode of deformation, and is additive for combined deformation, provided true plastic strain is taken as the measure of cold working.

The complex distribution of yield strength developed by weld shrinkage in previously reamed Inconel 600 sleeve is shown in Figure 6 (22). The line broadening data, converted to percent cold work and then yield strength, reveal a complex layer of highly cold worked surface material extending to a depth of 0.25mm in the reamed area adjacent to the heat affected zone. The plastic deformation caused by weld shrinkage extends 25mm to either side of the weld. The material is only fully annealed well beneath the reamed surface in the heat affected zone.

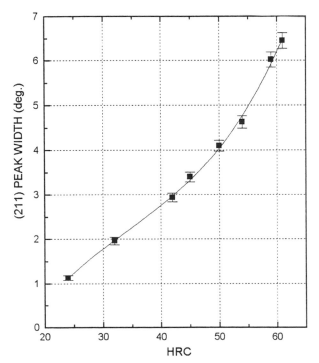

Fig. 3 - Dependence of (211) peak half-width on hardness for SAE 1552 steel. Data points are an average of five measurements using $CrK\alpha$, peak at $2\theta = 156$ deg. Ref. (16)

Stress corrosion cracks were associated with peak tensile stresses occurring just at the edge of the highly cold worked reamed area. Note that the yield strength of the deformed surface layers after cold working exceeds twice bulk yield of the alloy.

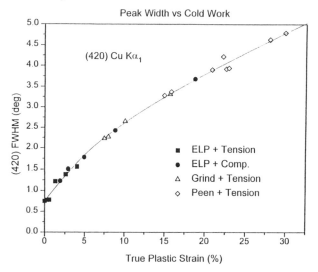

Fig. 5 - *Dependence of the (420) Kα_1 peak width on cold work (true plastic strain) showing independence of the mode of deformation and accumulation. Ref. (20)*

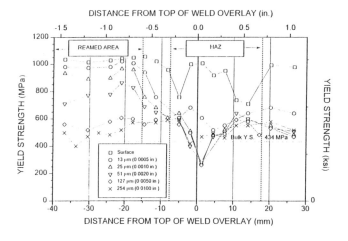

Fig. 6 - *Yield strength distribution calculated from peak breadth on the inside surface of Inconel 600 offer reaming and welding (see text). Ref. (22)*

MODELING OF RESIDUAL STRESS RELAXATION

The relaxation of residual stress during cyclic loading or at elevated temperature has been reported for decades, and has been reviewed (23). Recently, O. Vöhringer and coworkers (24,25) have developed models for the prediction of residual stress relaxation as functions of time and temperature, single cycle overload, and cyclic loading which promise to be powerful tools for failure analysis.

An Avrami approach is used to describe the fraction of residual stress remaining as a function of time and temperature in terms of an activation energy and to other material constants. The material dependent constants are developed from measurements of the isothermal stress relaxation as functions of time. The predicted and measured stress relaxation at the surface of shot peened AISI 4140 steel, using an incremental relaxation approach, is shown in Figure 7. The thermal relaxation model promises prediction of the retention of compressive residual stresses from shot peening in high temperature applications such as high performance gearing and turbine engine components for both failure analysis and design.

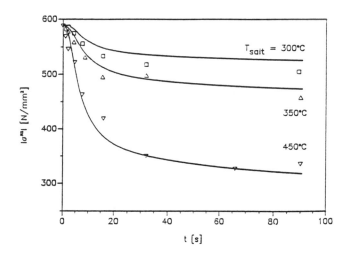

Fig. 7 - *Measured residual macro-stress after short time annealing in salt baths of different temperatures and relaxation curves modeled using the numerical stress-transient method based on an Avrami approach. Ref. (25)*

Momentary overload is commonly observed to upset and alter the state of residual stresses. Vöhringer has developed a model allowing prediction of the change in residual stress at the surface of a component as a result of plastic deformation. The residual stress and yield strength of the material in the current state and in each layer beneath the surface is incorporated into a finite element model allowing prediction of changes in surface residual stress. An example showing the change in the surface axial residual stress on 4140 shot peened steel in different heat treatments is shown in Figure 8. The model allows prediction of residual stress redistribution by subsequent deformation as in split-sleeve cold-expansion of reamed holes, overload of turbine disk bores at high RPM, and compressive overloading of shot peened components.

Cyclic loading causes residual stress relaxation for alternating stresses significantly above the endurance limit. Vöhringer has proposed a model describing the fraction of the initial residual stress remaining on the surface of a part exposed to cyclic loading as a linear function of log N, where the slope and intercept can be described by material dependent constants, which depend upon the stress amplitude. The model has been applied to both fully reversed bending and axial loading fatigue. The relaxation of surface axial residual stress in shot peened 4140 steel as a function of cycles is shown in Figure 9. After redistribution of stress on initial loading, the surface stress follows a linear reduction with log N, until near failure.

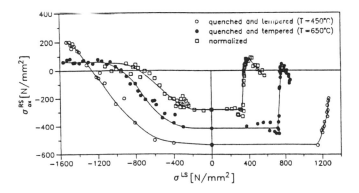

Fig. 8 - Axial macro-residual stress in differently heat treated and shot peened samples as a function of the quasi-static loading stress. Ref. (25)

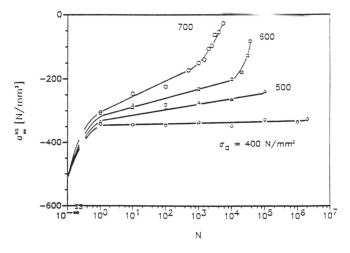

Fig. 9 - Axial macro residual stress of quenched and tempered samples as a function of the number of loading cycles for various applied stress amplitudes. Ref (25)

In failure analysis, the surface residual stress could be compared to areas of the specimen which were in a comparable state of residual stress prior to cyclic loading in order to estimate either the effective cyclic load or the number of cycles of exposure, if the other is known. Applications include prediction of relaxation of residual stresses under known cyclic loading in design and failure analysis.

CONCLUSIONS

A brief theoretical development and discussion of sources of error shows that the plane-stress model of x-ray diffraction residual stress measurement is the practical approach for engineering applications such as failure analysis and process development. Nondestructive surface residual stress measurements are inadequate for most applications because of errors inherent in uncorrected surface measurements, lack of correlation between surface stresses and the processes which produce them, and the need to know the subsurface peak residual stress to determine the effect on fatigue life.

The diffraction peak width obtained during residual stress measurement can be used to predict material properties such as hardness, percent cold work, and yield strength with high spatial and depth resolution. Current applications include detection of surface deformation producing softening of steels, measuring case depth with stress in induction hardening and increases in yield strength of machined surfaces of work hardenable alloys.

Recent developments in the prediction of residual stress relaxation using x-ray diffraction data have been successfully applied to predict thermal, cyclic, and single cycle upset relaxation of shot peened steels, and promise to become increasingly important tools in design, failure analysis, and process development studies.

REFERENCES

1   Koistinen, D.P. and R.E. Marburger, Trans. ASM, 51, 537 (1959)
2   Ogilvey, R.E., M.S. Thesis, MIT, 1952.
3   Hilley, M.E., ed., "Residual Stress Measurement by X-Ray Diffraction," SAE J784a, 21-24, Soc. of Auto. Engrs. (1971)
4   ASTM E837 "Standard Test Method for Determining Residual Stresses by the Hole-Drilling Strain-Gage Method," 1994.
5   B.D. Cullity, "Elements of X-ray Diffraction," 2nd ed., pp 447-76, Addison-Wesley, Reading, Massachusetts (1978)
6   Prevey, P. S., Adv. X-Ray Anal., 20, 345 (1977)
7   Dölle, H. and J.B. Cohen, Met. Trans., IIA, 159 (1980)
8   Dölle, H., J. Appl. Cryst., 12, 498 (1979)
9   Prevey, P.S. and P.W. Mason, "Practical Applications of Residual Stress Technology," ed. C. Ruud, pp 77-81, Am. Soc. for Met., Materials Park, Ohio (1991)

10   Ruud, C.O. and K. Kozaczek, "Proceedings of the 1994 SEM Spring Conference and Exhibits," pp 8-13, Soc. for Exp. Mechanics, Bethel, Connecticut (1994)

11   Noyan, I.C. and J.B. Cohen, "Residual Stress: Measurement by Diffraction and Interpretation," Springer-Verlag, New York, New York (1987)

12   Prevey, P.S., Adv. X-Ray Anal., 29, 103-111 (1986)

13   Gupta, S.K. and B.D. Cullity, Adv. X-Ray Anal., 23, 333 (1980)

14   Prevey, P.S. "Metals Handbook: Ninth Edition," Vol. 10, ed. K. Mills, pp 380-392, Am. Soc. for Met., Metals Park, Ohio (1986)

15   Moore, M.G. and W.P. Evans, Trans. SAE, 66, 340 (1958)

16   Hornbach, D.J., Prevey, P.S. and P.W. Mason, in Proceedings Conference on Induction Hardened Gears and Critical Components, Indianapolis, Indiana, May 1995, in press.

17   Prevey, P.S., "Practical Applications of Residual Stress Technology," ed. C. Ruud, pp 47-54, Am. Soc. for Met., Materials Park, Ohio (1991)

18   Koster, W.P., et al. "Surface Integrity of Machined Structural Components", AFML-TR-70-11, Air Force Materials Laboratory, WPAFB (1970)

19   Warren, B.E. and B.L. Averbach, J. of Appl. Phys., 23, 497 (1952).

20   Prevey, P.S., "Residual Stress in Design, Process and Materials Selection," ed. W.B. Young, pp 11-19, Am. Soc. for Met., Metals Park, Ohio (1987)

21   Prevey, P.S., " Workshop Proceedings: U-Bend Tube Cracking in Steam Generators," pp 12-3 to 12-19, Electric Power Research Institute, Palo Alto, California (1981)

22   Prevey, P.S., Mason, P.W., Hornbach, D.J., and J.P. Molkenthin, Paper presented at Materials Week, October 1994, Rosemont, Illinois, in press.

23   James, M.R., "Residual Stress and Stress Relaxation," E.Kula and V.Weiss, ed., pp 349-65, Plenum, New York, New York (1985)

24   Vöhringer, O., "Residual Stresses," E. Macherauch & V. Hauk, ed., pp 47-80, DGM Informationsgasellschaft-Verlag, Oberursel (1986)

25   B. Eigenmann, V. Schulze and O. Vöhringer, Proc. Int. Conf. on Residual Stress, ICRS IV, 598-607, SEM, (1994).

Compositional Mapping of Large Samples Using X-Ray Fluorescence

J. S. Krafcik and J. A. Brooks
Sandia National Laboratory, Livermore, CA

Abstract

An Energy Dispersive X-ray Fluorescence (EDXRF) Macroanalyzer was designed with the goal primarily to measure compositional variations in large metal samples. The analyzer consists of a Kevex manufactured X-ray fluorescence tube and detector, an automated X-Y stage, and a computer for control, graphical display, data processing and storage. The system can automatically raster samples weighing up to 136 Kg over an area 91 x 91 cm^2. The X-ray beam size can be varied in diameter from 50 microns to 13 mm by the use of different size collimators. The development and capabilities of this system are described and demonstrated through the characterization of compositional variations ranging from the macro to microscale in a 53 cm diameter, 3600 Kg ingot of Alloy 718.

Introduction

The degree of compositional variations in metallic components produced by casting or welding can be extremely important in determining part performance. These compositional variations can occur over the size scale of the dendrite spacing, on the order of microns, which is referred to as microsegregation. However, compositional variations can also occur over a much larger distance and is referred to as macrosegregation [1]. This bulk macrosegregation can, for example, be in the form of compositional variations from the edge to the center of the casting. Macrosegregation can also be more localized, often on the size scale of millimeters in the form of macrosegregation defects. Both types of macrosegregation are of special importance to producers and users of large ingots and castings since mechanical properties and performance can be degraded.

A schematic of the Vacuum Arc Remelting (VAR) process demonstrating the origin of these compositional variations is shown in Figure 1. With this process an electrode is remelted under vacuum into a water cooled copper crucible. Ingot sizes routinely melted range from ~20 to over 100 cm in diameter. During solidification, alloying partitioning occurs between the solid and liquid in the two phase region resulting in compositional variations between the first and last material to solidify, see Figure 1. This is the compositional variation which is referred to as microsegregation which occurs over the distances if the dendritic structure and is dependent upon alloy composition and process parameters. Increasing solidification rate reduces the size scale of the dendritic structure. Interdendritic liquid can also flow or be swept from the mushy zone and transported from one region of the ingot to another resulting in bulk macrosegregation, or be transported more locally within the two phase solid + liquid region itself resulting in macrosegregation defects. In general, the magnitude of the compositional variations exhibited with microsegregation are

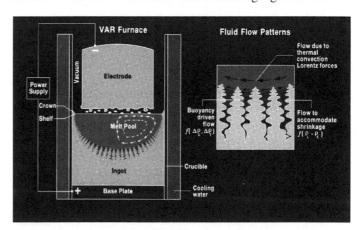

Figure 1. Schematic of vacuum arc remelting process showing the origin of microsegregation on the dendritic size scale and the transport of interdendritic fluid to different regions of the ingot resulting in macrosegregation.

much larger than those associated with macrosegregation.

The ability of characterizing compositional variations in large ingot structures is extremely desirable in the goal to improve ingot quality. Similarly, the ability to characterize compositional variations in critical components is also desired to improve component reliability or performance. However, no techniques have existed in which this could be easily accomplished. For many years the electron microprobe has proven to be a valuable instrument in determining microsegregation, but no similar system was available to analyze large scale macrosegregation, primarily because of size limitations on both the vacuum chamber and the electron beam.

In 1985, an out of vacuum X-ray Micro-analyzer was developed by Nichols and Ryon [2] that measures chemical composition on a size scale similar to that of an electron microprobe, with an analysis spot size of 10 to 100 μm. In 1986, Nichols, Boehme, Wherry and Cross developed the first prototype commercial system [3].

In 1989, an X-ray Fluorescence (XRF) Macroprobe with a total automated stage motion of 10 cm x 10 cm was developed for compositional mapping intermediate size samples [4]. An XRF system (Omicron - trademark of Kevex Corporation) is being marketed that is much more suited for measuring variations in composition. This system can analyze samples with an area up to 15 x 24 cm without readjusting the sample. The system is automated with an X-Y stage that is integrated with the analysis system so that composition can

be mapped over the sample surface. The spot size can be varied from 50 μm to 1 cm. Again, this unit was designed for analysis of intermediate size samples weighing up to 4.5 Kg.

A new system with an X-ray beam size range from 50 microns to 13 mm has been developed that has expanded the analysis capabilities over an 91 x 91 cm^2 area for samples weighing up to 136 Kg. The capabilities of this system and examples of sample analysis of ingots are discussed. However, it should also be noted that the system can be used for applications other than compositional analysis, for example to determine the uniformity of metallic coatings [5].

Instrument Design and Operation

The Macroanalyzer consists of a modified Kevex Omicron Mo-targeted micro focus X-ray tube, SiLi energy dispersive detector, a computer automated X-Y stage, a table to hold and level the sample, a housing for structural support, X-ray shielding and a microcomputer with software [6]. A schematic of the system is shown in Figure 2.

The X-ray beam size is set at the source using collimators between 50 microns to 13 mm which are physically placed at the X-ray tube to narrow the beam to a desired spot size [7]. This is done prior to placing the sample for analysis. The X-ray source has a maximum energy of 50 kV, 1.0 mA. In the analysis shown below a 30 kV, 0.09 ma beam was used. After setting the beam size, the sample to be analyzed is placed on the stage and moved under the source and detector.

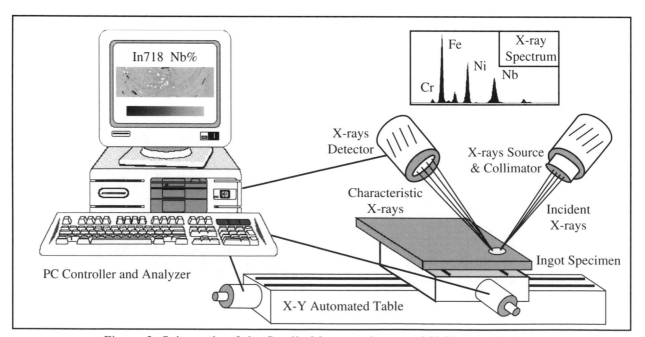

Figure 2. Schematic of the Sandia Macroanalyzer and X-Y controlled stage.

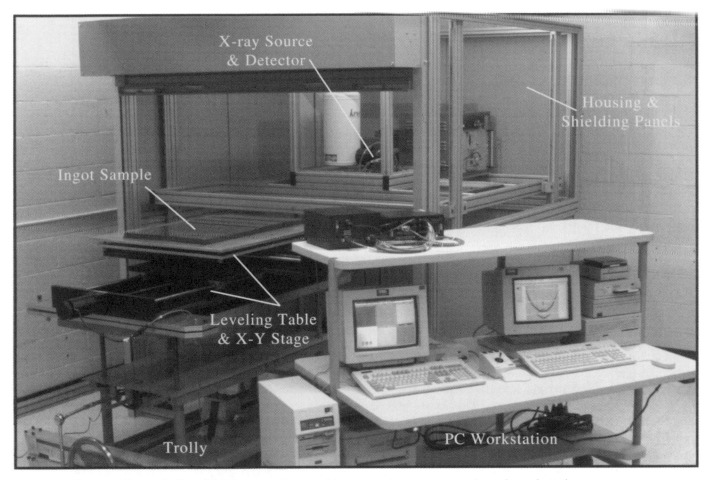

Figure 3. Sandia Macroanalyzer with supporting components and workstation.

A 2 x 2 x 2 meter enclosure houses the sample stage, detector and source as shown in Figure 3. The enclosing frame is anchored to the floor and wall to minimize vibrations and to structurally support the stage. Sliding steel panels are placed in tracks of the frame for X-ray shielding. The panels are interlocked to safely power down and disconnect the source in the event of exposure. The stage, which supports the sample, includes a trolly and an X-Y positioning table that covers a 91 x 91 cm^2 area. The stage and trolly can support samples weighing up to ~136 Kg. The trolly rolls on a track which allows the operator to place and position a sample or samples on the stage outside the housing. After the sample is positioned and leveled on the stage, it is pushed into the housing under the source and detector. Here, locating the sample in the housing is done through a NTSC video camera which views at the focal point of the X-ray source. The camera's image field of view is 3 x 4 mm^2 with a graticule at the center of the screen for point location. A joystick helps the operator manipulate the X-Y table to the precise analysis location. When the sample is within the housing, all the doors and panels are closed and interlocked before an analysis can begin.

All operations are controlled outside the housing at the workstation, Figure 3. This computer controls the X-Y stage rastering, X-ray source voltage and current, safety interlocks, spectrometer acquisition, and final data acquisition and mapping. Rastering the X-Y stage to each pixel location is done by the operator entering the total X and Y distance and the step size in the X and Y direction. The spectrometry program, Kevex's Toolbox®, is coupled with the X-Y stage communication procedure for automatic rastering. Upon completing a spectral acquire, deconvolution, and quantifying the data, the X-Y procedure rasters the sample to a new pixel location. At completion of the entire scan an ASCI file is generated with each pixel having an x, y position, spectral counts for each element identified and in our examples quantified for weight percent of each element.

For detection of macrosegregation defects, it is desirable to reduce the X-ray spot to the same size scale as that of the defect. By optimizing the spot size and counting time, the analysis can be performed with reasonable speed and sensitivity. For example, to

113

analyze for niobium in Inconel 718, a 500 μm spot size at the sample's surface requires 200 seconds per step for analysis, while a 2 mm spot requires a counting time of 60 seconds to achieve the same total counts required for analysis.

Data mapping was done with Spyglass Transform, which imports and renders the data into a color spectral map with corresponding color bars, labels and dimensions. A map is generated by importing data into a spreadsheet and rendering a grid with each pixel representing a data point. Color scales are placed to identify minimum to maximum data. A smooth contour map is created by putting the data through an interpolation routine and rendering. It was noted that the commercial market has many software packages for 2 & 3D renderings on MAC, SUN, PC, and other machines, which will work independently with our data format and with good output.

Compositional Analysis

One of the goals of the macrosegregation analysis is to characterize the nature and degree of macrosegregation and to provide insight into the formation mechanisms of macrosegregation defects. This knowledge is then used to verify computational models, and related to melt conditions. Here we will concentrate primarily on demonstrating the capabilities of the system in the characterization of large ingot structures.

A 53 cm diameter 216 cm long VAR ingot of Alloy 718 with the nominal alloy compositions given in Table 1, was cut into sections ~ 18 cm long [8]. These sections were then sliced longitudinally and a slab 2 - 3 cm thick was taken from the center of each section to produce 11 slices comprising the ingot length.

A Nb compositional map of the total ingot is shown in Figure 4. This data was obtained using a grid size of 13 x 13 mm and a beam diameter of 13 mm. The ingot was melted under a variety of conditions to obtain correlations between melt conditions, pool shape, and defect formation and thus deviated from normal melt practices. It is evident in Figure 4 that extensive bulk macrosegregation exists. The measured composition ranges from ~ 4.9 to 6.1 % Nb (all compositions reported as wt%) when using a beam size of 13 mm. The extreme minimum value in regions near the top of the ingot are anomalies associated with shrinkage

Table 1. Composition of Inconel 718 casting in weight %.

Ni 53.55	Cr 17.69	Fe 18.44
Nb 5.34	Mo 3.02	Ti 1.02
Al 0.53	Mn 0.07	Si 0.18

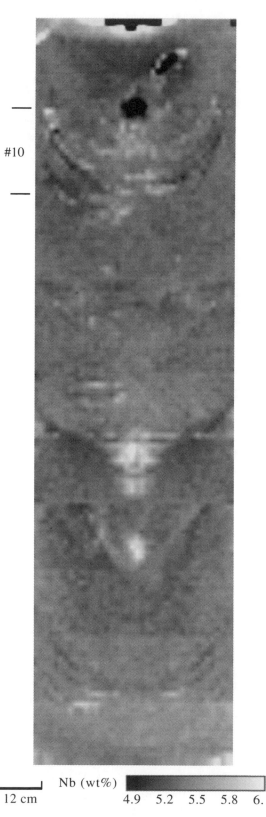

Nb (wt%) 4.9 5.2 5.5 5.8 6.1

12 cm

Figure 4. Compositional map of Nb taken of the entire Alloy 718 ingot showing the nature of macrosegregation. Data taken with a 13 mm beam size and 13 mm grid spacing.

114

pipes. It should also be pointed out that analysis was conducted on unetched ingot slices except for the bottom two sections. The somewhat higher than nominal Nb content of these slices are associated with preferential etching of phases resulting in compositional changes of the surface region. In some regions of the compositional map, the outline of the pool can be estimated and solute dumps [9] associated with melt changes are evident.

Figure 5 shows the correlation between the ingot structure and the digital XRF map of niobium in a single ingot slice noted as #10 near the top of the compositional map in Figure 4. The most dominate features of the ingot macrostructure shown in Figure 5a are freckles near the mid-radius and upper center which are known to be Nb enriched channel type macrosegregation defects [10]. A Nb map of the same slice is shown in Figure 5b. Data for this map was taken in the unetched condition using a beam size of 6.4 mm and a grid spacing of 6.4 x 6.4 mm resulting in a grid of 27 x

82 pixels. The higher spacial resolution resulting form the finer spot size and grid spacing is apparent when comparing Figure 5a to Figure 4. This is especially apparent when examining the freckle regions in which not only are the higher Nb regions associated with the freckles are more visible, but a Nb depleted region is can be seen to exist directly above the freckles. The Nb distribution in the freckle region is consistent with the microstructure of the freckle shown in Figure 6. The freckles are comprised largely of Nb containing intermetallics (primarily M_2Nb). The region directly above the freckle contains less of these second phase particles. In the analysis using a 6.4 mm diameter beam, the variation in measured minimum to maximum Nb content was increased to ~2 wt. %.

The region of the freckle was further analyzed using a beam size of 500 microns in an Omicron in X-Map mode. The yet higher increase in both spacial and compositional resolution is apparent in both the Nb and Cr maps shown in Figure 7a and b. An analysis across

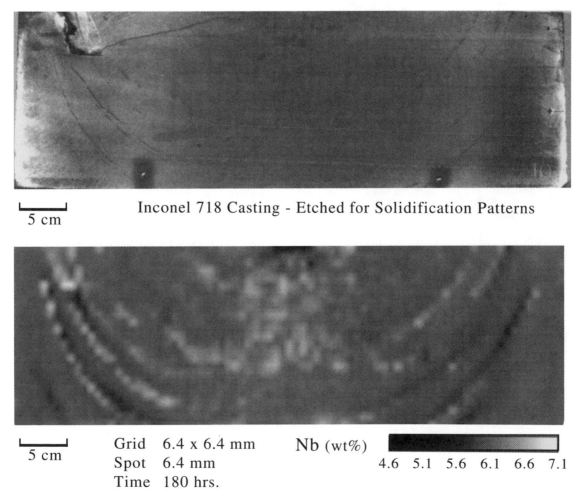

5 cm · Inconel 718 Casting - Etched for Solidification Patterns

5 cm

Grid 6.4 x 6.4 mm
Spot 6.4 mm
Time 180 hrs.

Nb (wt%)

4.6 5.1 5.6 6.1 6.6 7.1

Figure 5. (a) Macroetched ingot slice showing regions of freckle microsegregation defects. (b) compositional map for Nb of slice in (a) taken with a 6.4 mm beam size and 6.4 cm grid spacing.

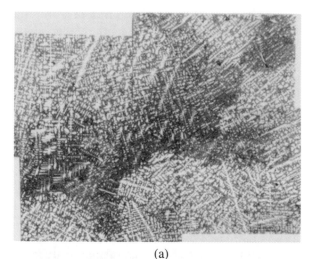

(a)

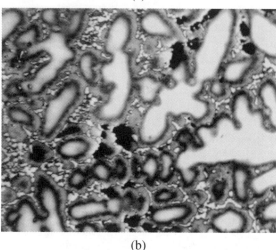

(b)

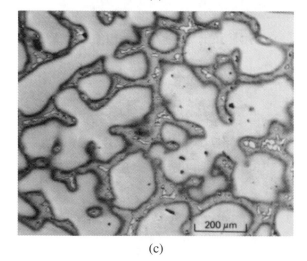

(c)

Figure 6. (a) Montage of freckle showing nature of dendritic structure and porosity associated with freckle. (b) higher magnification of freckle region showing high density of Nb containing solidification products within freckle (c) region directly above freckle with depletion of secondary solidification products (low Nb concentration).

one of the freckles is shown in Figure 7c. It can be seen that the 500 micron beam size is considerably smaller than the size of the microsegregation defect, but from the small amount of scatter in the data in the region of the freckle, is large compared to the size scale of the individual phases. The variation in Nb content is now seen to be ~6 wt. %. The level of resolution is probable near the limit of interest in developing an understanding of the defect formation. When the spacial resolution is further increased compositional variations on the level of microsegregation would be detected, and in this case microprobe would be an alternate analytical technique. However in this case a metallographic size sample would need to be removed from the ingot slice.

In some cases it is possible to detect segregation on the dendritic size scale (microsegregation) even with a beam size as large as 500 microns. This, of course, dependents upon the size scale of the dendritic structure. An example is shown in Figure 8 of a region near the center of the ingot in which the dendritic size scale is fairly large due to increased local solidification time. The region analyzed is shown in Figure 8b in which the light etching regions correspond to Nb lean intradendritic cores and the dark regions to Nb enriched interdendritic boundaries. The large light etching feature in Figure 8b is encompassed within a single dendritic grain. These metallographic features on a microscale are apparent in the Nb qualitative map shown in Figure 8c where the Nb lean primary and secondary dendrite cores are dark and the Nb enriched interdendritic regions are white.

Summary

An XRF system has been developed to compositionally characterize large samples measuring up to 91 x 91 cm^2 and weighting up to 136 Kg. The beam size can be varied from 50 microns to 1 cm with the use of collimators placed at the X-ray tube.

This system provides the capability of compositionally mapping large samples and further interrogating regions of interest at finer and finer size scale. Examples were shown where large samples were mapped for compositional variations over the total ingot surface (macrosegregation) and specific regions analyzed on the size scale of the dendrite structure (microsegregation). A trade off exists between the size of the area analyzed, pixel grid spacing, and beam size and intensity with that of analysis time.

The area analyzed and resolution required are dependent upon the features of interest.

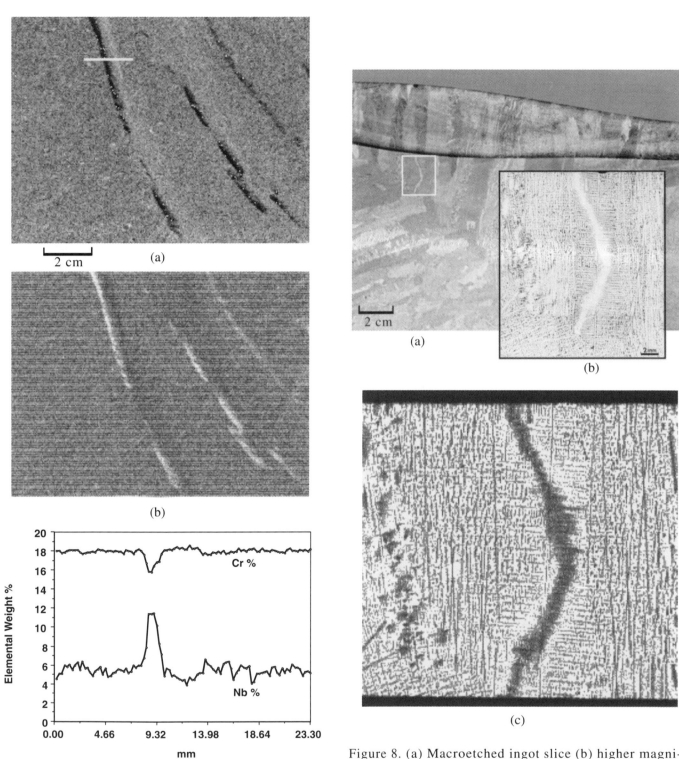

Figure 7. Cr and Nb maps taken in the region of a freckle from ingot slice shown in Figure 5 taken with a 500 micron collimator. Freckle is depleted in Cr (a) and enriched in Nb (b). Note region above freckle is enriched in Cr and depleted in Nb. (c) line scan across freckle indicated in (a), showing quantitative analysis.

Figure 8. (a) Macroetched ingot slice (b) higher magnification of region of ingot analyzed showing dendritic structure. (c) Nb map of region in (b) showing micro-segregation consisting of Nb dendritic cores and Nb enriched interdendritic boundaries.

117

Acknowledgments

We acknowledge Dale Boehme of Sandia National Laboratories for Omicron analysis and his critical review of the manuscript. This work was supported by the U. S. Department of Energy under Contract Number DE-AC04-76DP00789, and the Specialty Metals Processing Consortium, Contract Number DE TT04-90AL 65451.

References

1. M. C. Flemings: Solidification Processing, McGraw Hill (1974).

2. M. C. Nichols, R. W. Ryon: An X-ray Microfluorescence Analysis System With Diffraction Capabilities, *Advances In X-ray Analysis*, Charles S. Barret et al., ed. Plenum Press, N.Y., Vol. 29 (1986).

3. D. Wherry and B. Cross: XRF, Microbeam Analysis, and Digital Imaging Combined into Powerful New Technique, *Kevex Analyst*, 12:8, Kevex Corporation, Foster City, California 904404 (1986).

4. N. L. Gelfrich, D. E. Leyden, and E. A. Erslev: XRF Macroprobe Analysis of Geologic Materials, *Advances In X-ray Analysis*, Plenum Press, N.Y., Vol 33, pp.593-601 (1990).

5. E. P. Bertin: Principles and Practice of X-ray Spectrometric Analysis, 2nd Ed. Plenum Press, N.Y., pp.811-828 (1975).

6. J. S. Krafcik and J. A. Brooks: "Development of an Macro-Analyzer for Compositional Mapping Macrosegregation of Large Ingot Samples", report to SMPC, Sandia National Laboratories, Jan. 1993.

7. D. A. Carpenter: *X-Ray Spectrometry,* Vol. 18, pp.253 (1989).

8. J. A. Brooks and J. S. Krafcik: "Metallurgical Analysis of a 520 mm Diameter Inconel 718 VAR Ingot", report to SMPC, Sandia National Laboratories, Aug. 1993.

9. M. C. Flemings, R. Mehrabian and G. E. Nereo: Trans TMS-AIME, 1968, vol. 242, pp.41-49.

10. J. A. Domingue, K. O. Yu, and H. D. Flanders: Characterization of Macrosegregation in ESR IN-718, in Fundamentals of Alloy Solidification Applied to Industrial Processes, symp. NASA Lewis, Sept. 1984.

Recent Advances in the Field of Scanning Auger Microscopy (SAM)

Brian R. Strohmeier
ALCOA Technical Center, Alcoa Center, PA

Abstract

During the past 20 years Auger electron spectroscopy (AES) has become a popular and widely used analytical technique in industry for characterizing the surface composition of solid materials. Scanning Auger microscopy (SAM), the technique of imaging the relative elemental concentration on the top few atomic layers of a solid by measuring the emitted Auger electrons over a given sample area, is distinguished among surface analysis techniques by its unique combination of surface sensitivity (i.e., 20-30 Å); the ability for depth profiling analysis when combined with ion sputtering; high spatial resolution (i.e., <500 Å); relative ease of identifying and quantifying most elements (i.e., ≥ Li), secondary electron imaging and surface elemental mapping capabilities; and the ability to distinguish between different chemical states for many elements. Within the past few years, two major advances have been made in commercially available Auger instrumentation: the development of Schottky (thermally-assisted) field emission (SE) electron sources and multi-channel electron detectors (MCDs). Field emission electron sources can produce optimum primary beam diameters on the order of 150 Å or less, providing much higher spatial resolution than conventional tungsten filament or lanthanum hexaboride (LaB_6) cathode sources. Field emission sources also provide higher beam current densities than conventional sources, which increases the instrument sensitivity (counting rate). Multi-channel detectors also provide increased instrument sensitivity, allowing faster data acquisition and reduced sample damage due to radiation effects. Auger instruments that combine these two features allow many complex analyses, which were beyond the capabilities of previous Auger instruments, to be performed routinely.

AUGER ELECTRON SPECTROSCOPY (AES) is based on the radiationless, intra-atomic electron transition effect first observed in 1923 by Pierre Auger, who was studying the photoelectric effect in gases using a cloud chamber (1). Auger found that after the initial production of a photoelectron, an additional electron was emitted from the same position as the preceding photoelectron. He soon discovered that the energy of these secondary electrons was independent of the energy of the primary X-ray source and that the electron energy was characteristic for different elements (2). These secondary electrons, now named after Auger, are produced whenever incident radiation (i.e., photons, electrons, ions or neutral atoms) interact with an atom with an energy exceeding that necessary to remove an inner-shell (i.e., K, L, M, ...) electron from the atom. This interaction leaves the atom in an excited state with a missing inner-shell electron or core hole. These excited atoms are unstable, and de-excitation occurs immediately, by either emission of an X-ray photon (a fluorescence transition) or emission of an Auger electron (the Auger effect).

The Auger effect forms the physical basis of Auger electron spectroscopy, which did not become a commercially viable technique until 1967 when ultra-high vacuum equipment and the necessary electron energy analyzers and detection systems were developed and readily available (3). Since that time, AES has become one of the most widely used surface characterization tools. It is able to provide elemental composition (except H and He) and quantitative information about the topmost atomic layers of solid materials. The relatively low kinetic energy (i.e., ~50-2000 eV) of the emitted Auger electrons is the main reason for the surface sensitivity of the technique, since the corresponding electron mean free path in solids is on the order of several atomic distances. The kinetic energy of an Auger electron depends on the binding energies of

119

the electrons in the involved shells of the excited atom and, therefore, carries the characteristic elemental and chemical state information. From the relative intensities of the individual Auger electron signals, the elemental concentrations in the surface layer can be quantified with a proper knowledge of the experimental parameters and/or the use of reference materials. The technique of scanning Auger microscopy (SAM), which involves the imaging of the relative elemental concentration on the top few atomic layers of a solid by measuring the emitted Auger electron signal over a given sample area, has also become a major problem-solving technique in many industrial laboratories.

A detailed treatment of the basic principles, instrumentation and many practical applications of AES and SAM are far beyond the scope of this article, but is available in many recent textbooks and review articles (4-10). The focus of this paper is on two major advances that have occurred in commercially available Auger instrumentation within the past few years: the development of high spatial resolution field emission electron sources and multi-channel electron detectors (MCDs).

As an analytical technique, SAM is unique in that its imaging capability is sandwiched between X-ray photoelectron spectroscopy (XPS), which has the same surface sensitivity, but poorer spatial resolution (i.e., 10-100 μm), and electron probe microanalysis (EPMA), which has similar spatial resolution, but poorer surface sensitivity (i.e., 1 μm) (11). Therefore, SAM is most useful for those problems which require both good depth resolution and high spatial resolution. In general, a SAM instrument consists primarily of an ultra-high vacuum scanning electron microscope (SEM) column producing a finely focused primary electron beam and an electron energy analyzer, which is capable of scanning a narrow kinetic energy window in front of an electron detector system. High spatial resolution Auger imaging is generally performed by focusing the primary electron beam into a fine spot and subsequently positioning it at a known pixel location (6). The electron energy analyzer is then set to sequentially monitor each of the specific kinetic energy windows of the elements of interest at each pixel location as the primary beam is scanned across the desired area of the sample surface. Any signals detected in this manner are ascribed to the appropriate individual pixel coordinates.

For the highest possible spatial resolution, it is beneficial to maximize the current in the sample and to minimize the primary electron beam diameter, within the limits imposed by sample charging and the possibility of radiation damage. The tungsten filament and lanthanum hexaboride (LaB$_6$) cathode electron sources typically available in SAM instruments have usually provided a trade-off between good spatial resolution and high beam currents. However, recently developed field emission

electron sources using Schottky (thermally-assisted) emitters provide an intense electron source while retaining good beam current stability (6,12,13).

Because of the typically weak signals encountered in AES, one usually measures no more than two energy channels (i.e., peak and background) per pixel (for each element) to obtain the imaging signal in a reasonable time frame. Other restrictions to the acquisition time required to obtain high quality images can occur due to instrumental drifts and radiation damage. The recent development of multi-channel detection systems for SAM instruments has greatly improved the instrumental sensitivity by allowing simultaneous collection of several energy channels, which can be summed and signal averaged by computer (14). This increased sensitivity provides much faster data acquisition times and reduced sample damage due to radiation effects.

The recent advances made in field emission electron sources and multi-channel detectors for SAM instrumentation will be described in greater detail below.

Schottky Field Emission Electron Sources and Associated Electron Optics

Two types of electron sources are used in commercially available SAM, SEM and related electron beam instruments: thermionic and field emission electron cathodes (15-19). Thermionic cathodes emit electrons from the cathode material when they are heated, whereas field emission cathodes rely on a high electric field to draw electrons from the cathode material by electron tunneling mechanisms (15-17).

Tungsten filaments were one of the first thermionic electron sources developed. They have a relatively high work function (about 4.5 eV), which means they require a large amount of heat to emit a given amount of electrons and consequently, they have a relatively low brightness compared to other sources. Tungsten filaments have moderate vacuum requirements (<10^{-5} torr) and short lifetimes (about 100 hr); however, they are relatively inexpensive compared to other sources.

LaB$_6$ cathodes were developed to provide higher brightness than tungsten sources. LaB$_6$ cathodes are more efficient emitters than tungsten because they have a lower work function (about 2.6-2.7 eV) and, therefore, can operate at lower temperatures with resulting longer lifetimes (15,17). LaB$_6$ filaments are more costly than tungsten, typically ten times; however, the increased current in a given beam size and the significantly longer lifetime (≥1000 hr) justify the increased cost for many applications (17).

Field emission cathodes were developed to provide higher resolution (smaller beam size), higher brightness (current) sources compared to thermionic sources. Two types of field emission sources are available: cold field emission (CFE) cathodes, which are made from <310>,

<111> or slightly oxidized <100> oriented tungsten wire and Schottky (thermally-assisted) field emission (SE) cathodes made from ZrO coated <100> tungsten wire (15,18). The basic underlying mechanism for generating electron beams from these two sources involves emission of electrons from a metal surface under the influence of a strong electric field by electron tunneling mechanisms (16-18).

CFE cathodes operate at room temperature and electrons tunnel from various energies below the Fermi level. SE cathodes operate at about 1800 K. The ZrO coating lowers the tungsten work function from 4.5 eV to about 2.8 eV to provide high emission and good stability (17). Hence, with SE cathodes, thermally excited electronics escape over a field-lowered potential energy barrier (15-18). This high temperature stability allows the use of SE cathodes at pressures in the high 10^{-9} torr range instead of the low 10^{-10} torr range typically required for proper CFE operation.

The restrictive, ultra-high vacuum requirement for CFE sources is due to their high susceptibility to poisoning by adsorbed hydrocarbons or other contaminants (15). Such contamination alters the work function of the tip, causing reduced and unstable emission. Only heating in 10^{-6} torr of oxygen can remove the contamination. This process, known as "flashing," cleans off the adsorbed gas atoms. The high operating temperature of the SE cathode minimizes the effects of adsorbed contaminants; hence, the operating pressure can be several orders of magnitude higher than that required for the CFE sources without having a significant effect on the electron emission and beam stability (14,15). Beam stability is an important parameter for a SAM instrument because of the relatively long data acquisition times typically required for obtaining Auger images.

Table 1 summarizes the nominal operating characteristics and other features of the four main electron sources described above (13-18). The brightness of the source is a key parameter in determining the optical performance of the electron beam column. The brightness of the SE cathode is comparable to that of the CFE cathode and is two to three orders of magnitude greater than that of the two thermionic sources. Hence, for a given sample current, one needs to accept a much smaller angle in the electron optical system for the field emission sources compared to the thermionic sources. This serves to dramatically reduce the effects of the spherical and chromatic aberrations in the optical system (14). The energy spread of the emitted electronics also affects the chromatic aberrations and the coherence of the beam (14,16). In the case of the SE cathode, the energy spread is about 1 eV, which is somewhat greater than that of the CFE source, but comparable to the thermionic sources.

The smaller the virtual source size, the less demagnification is required by the electron optics. This reduces the number of required electron lens and, most importantly, allows a larger effective acceptance angle at the source for a given objective aperative diameter, which results in more current at the sample for a given focused beam size. The virtual source size is not the same size as the emitting surface (see Table 1), but is the diameter of the virtual image at the crossover point in the electron optics column. The virtual source size of the field emission cathodes is many orders of magnitude less than that of the thermionic sources, which ultimately leads to much smaller beam diameters and higher spatial resolution. The SE virtual source size is slightly larger than that of the CFE. Therefore, the SE source requires greater demagnification to produce an equivalent spot size. The two major advantages of the SE source

Table 1. Comparison of various electron source characteristics.

Nominal Operating Properties and Features	SE Cathode	CFE Cathode	LaB$_6$ Cathode	W Filament
Brightness (A/cm^2 sr)	5×10^8	10^9	10^7	10^6
Energy Spread (eV)	0.3 - 1.0	0.2 - 0.3	1.0	1.0
Emitting Surface Area (μm^2)	>0.3	0.03	>>1	>>1
Virtual Source (Crossover) Diameter (nm)	15	3	10^4	>10^4
Short-Term Beam Current Stability (% RMS)	<1	4 - 6	<1	<1
Operating Temperature (K)	1,800	300	1,400 - 2,000	2,800
Tip Flashing Required?	No	Yes	No	No
Required Operating Vacuum (torr)	<10^{-8}	<10^{-10}	<10^{-7}	<10^{-5}
Typical Service Life (hr)	5,000 - 10,000	2,000 - 10,000	200 - 1,000	40 - 100
Typical Cost ($U.S.)	$1,000 - 1,500	$500 - 2,000	$500 - 2,000	$20

compared to the CFE source are the greater emission stability and the less restrictive vacuum requirements. These advantages are only slightly offset by the larger demagnification required, which reduces the current available at the smallest beam sizes. The SE source also has small advantages over the CFE source when considering the nominal service lifetime and replacement cost (14,15,17). Hence, considering a composite of the properties listed in Table 1, the SE cathode is the optimum choice for an electron source in a scanning Auger instrument.

The first commercially available SE source for SAM instruments was introduced in late 1990 by Perkin Elmer, now Physical Electronics, Inc. (PHI), Eden Prairie, MN, in the "Model 670 Scanning Auger Nanoprobe" instrument (14,19). Since that time, similar SE sources have also become available from other SAM manufacturers such as Fisons Instruments, U.K., and Cameca Instruments, France (12,20).

A schematic of the PHI SE source is shown in Figure 1 (14). The SE cathode consists of a wire of single-crystal tungsten with a ZrO coating that is fashioned into a sharp point and spot welded to a tungsten hairpin wire (17). A very sharp tip radius (<1000 Å) is important so that the electric field can be concentrated to an extreme level. A filament supply provides the current which resistively heats the electron source to the nominal 1800 K operating temperature. A 300-volt source provides a reverse bias on the suppresser electrode, which prevents emission off of the shank of the emitter tip. An extraction power supply (V_{ext}) provides the voltage which stimulates emission off the tip of the emitter. Useful electron emission comes off only the very tip of the emitter and is collected by the extraction electrode. The gun focus voltage (V_{foc}) defines the focal length or operator to vary the specimen current at a fixed objective aperture setting.

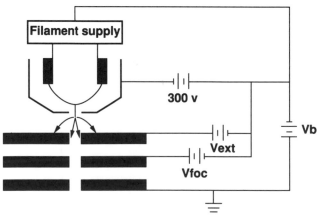

Fig. 1 - Thermal field emission electron source and electrostatic gun lens.

Figure 2 shows a cross-sectional diagram of the primary electron beam column in the PHI instrument,

which houses the field emission electron source (13,14). The column consists of an upper chamber and a lower chamber. The upper chamber, which houses the SE source, is differentially pumped by a 60 liter/sec ion pump on top of the column to keep the pressure near the emitter as low as possible. The SE source has a manual X/Y translation capability for alignment purposes, a three-element electrostatic lens (see Figure 1 for greater detail) and a set of electrostatic deflection plates (gun steering) for beam steering (13,14). An isolation valve can be used to isolate the upper gun chamber from the lower specimen chamber to keep the electron source under vacuum if the specimen chamber must be "brought up to air" for repairs to other components.

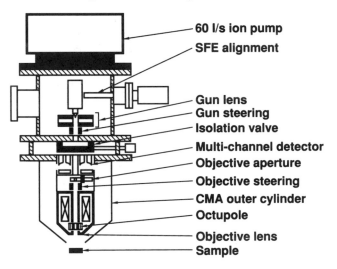

Fig. 2 - Cross-sectional diagram of the primary beam column showing the coaxial field emission electron source and the cylindrical mirror electron energy analyzer (CMA).

The lower chamber contains a magnetic objective lens, a second set of electrostatic deflection plates (objective steering), a variable objective aperture, an octupole, which is used for beam scanning, image shift and correction of astigmatism and finally, the multi-channel electron detection system, which is associated with the cylindrical mirror electron energy analyzer or CMA (14).

A simple schematic of how these electron optics operate is shown in Figure 3 (13,14). At the top of the diagram is the emitting source, which is initially focused by the electrostatic gun lens. The emitter tip produces a virtual source near the point of emission. That virtual source is focused to an intermediate crossover point by the electrostatic lens, at which point the beam illuminates the mechanical objective aperture. The size of the objective aperture, which can be easily altered by the operator, determines what portion of the beam current is transmitted to the sample. This image of the source is focused onto the sample by the magnetic objective lens.

The beam is electrically aligned through the column using the upper (gun steering) and lower (objective steering) electrostatic deflection plates. As stated above, the octupole, which is visible directly above the sample in Figure 3, is used for final beam scanning, image shift and correction of astigmatism.

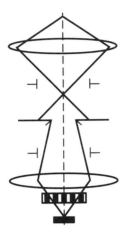

Fig. 3 - Schematic of field emission electron gun optics.

The optimum final electron beam size at the sample with this type of electron source and electron optics is on the order of 100-150 Å (12,13,14,20), which is a significant improvement over the 300-1000 Å optimum beam size typically found in SAM instruments with thermionic electron sources. In any case, at equivalent beam size, the SE source would produce significantly higher current at the specimen, and hence, higher sensitivity (counting rates) compared to the thermionic electron sources. In practical terms, this means that SAM instruments, which utilize SE electron sources, can now perform routine analyses easily and rapidly on sample features on the order of 1000 Å or less (21). Such capability has created numerous new applications in fields such as metallurgy, microelectronics failure analysis, semiconductor processing and advanced materials research (21). However, it must be pointed out that regardless of the optimum beam size at the specimen, the actual spatial resolution for both point and imaging analyses will be determined by backscattering effects in the sample itself (22). This is because the backscattered primary electron beam will also cause Auger emissions in the sample (22). In general, the extent of the backscattering effect increases with the average atomic number of the specimen (17,22). Operating at lower primary beam voltages helps reduce the backscattering limitations to spatial resolution, but also generally results in a larger primary beam diameter. However, with SE sources the trade-off between beam voltage and beam size is not as severe as that found for thermionic sources. It should be remembered that the optimum beam voltage for maximizing the spatial resolution depends on the instrument, the Auger

transition being measured and the sample itself (22). For general analytical work, the ideal SAM instrument should be flexible enough to operate over a wide range of beam voltages and currents while maintaining a small beam size. An SE electron source is more than capable of fulfilling this requirement (22).

Multi-Channel Detection (MCD) System

At the same time that PHI introduced the first commercial SE electron source, they also introduced the first (and to date only) commercial multi-channel detector in a CMA, a feat that many electron optics scientists considered impossible (14,19). This was a significant advancement in CMA technology because it allowed many energy channels to be measured simultaneously, resulting in a much higher signal. Since that time, Fisons Instruments, U.K., and JEOL Ltd., Japan, have introduced similar MCD systems in their own SAM instruments, which are both based on hemispherical sector electron energy analyzers or HSAs (12,23).

Details of the operating principles of CMAs are available elsewhere (4,7,8,10). Basically, a CMA consists of an inner and outer cylinder as shown in Figure 4 (14). The primary electron beam is coaxial with the CMA. The entrance silt to the CMA defines the acceptance angle of the analyzer for the emitted Auger electrons. A negative voltage is applied to the outer cylinder, which provides the radial electric field that performs the electrostatic deflection and energy dispersion function of the analyzer.

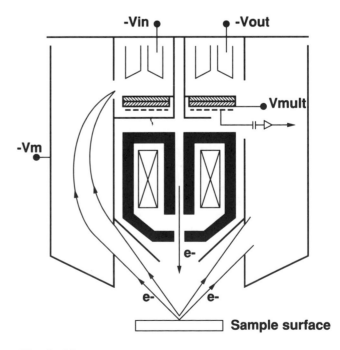

Fig. 4 - Multi-channel cylindrical mirror analyzer (CMA).

In a conventional single-channel CMA instrument, the detector is positioned inside of the inner cylinder at the axial position of the exit slit (7,8). The detector will detect only those Auger electrons that have a kinetic energy equivalent to the specific analyzer energy. By sweeping the voltage on the outer cylinder, one acquires the spectrum by detecting the distribution of electrons as a function of kinetic energy.

Figure 4 shows the CMA arrangement in the PHI multi-channel CMA system (14). The exit slit on the inner cylinder is opened such that a wider band of electron energies is passed by the CMA. Additional electrostatic elements ($-V_{in}$ and $-V_{out}$) within the MCD assembly project the focal surface of the CMA, not to a point at a single detector, but onto the face of a microchannel plate electron multiplier (cross-hatched region in Figure 4). The PHI system incorporates a concentric ring anode structure into the multi-channel plate electron multiplier, which allows electrons with eight different energies to be detected simultaneously (14). As the "bundle" of electrons enters the CMA exit slit, the highest energy electrons are deflected the most and are collected on the outer radius of the microchannel plate. Moving progressively to lower energies, the lowest energy electrons are deflected most weakly within the MCD assembly and are collected on the inner radius of the microchannel plate.

The microchannel plate consists of a thin glass wafer with numerous channels (~20 μm dia, 20-30 μm pitch) etched through the glass (14). The glass has a high secondary ion yield, and as electrons pass through the detector, the enhanced secondary electron yield results in electron multiplication. With this detection system, eight individual spectra can be collected in parallel. They can then be summed and fully integrated by computer for an approximate 8-fold increase in instrument sensitivity compared to a single-channel detector CMA instrument (14,19). This detection system results in significantly increased speed of analysis compared to earlier instruments.

Summary

Two recent instrumental developments have dramatically changed the field of Auger spectroscopy: the development of high brightness Schottky field emission sources for the primary electron beam and the development of multi-channel electron detectors in existing electron energy analyzers (both CMA and HSA). In broad terms, these two developments have resulted in significantly higher instrument sensitivity (counting rates) at higher spatial resolution (smaller beam sizes) with greatly reduced analysis times and reduced sample damage due to radiation effects.

SE electron sources have distinct advantages over traditional thermionic sources (i.e., W and LaB_6) in terms of brightness, energy spread, virtual source size, service life and ultimately, beam size and current density at the specimen. Although the spatial resolution of the CFE sources often used in SEM instrument is somewhat higher than that obtainable with SE sources, CFE sources have less beam stability and are susceptible to poisoning, resulting in more restrictive vacuum requirements compared to SE sources. Therefore, when one considers a composite of the most important properties of electron sources, the SE source is the optimum choice for an Auger instrument.

The development of MCDs for Auger spectrometers was a significant advancement, especially in CMA technology, because it allows many channels of energy to be measured simultaneously and summed, which results in much higher counting rates. This feature increases analysis speed significantly and helps to simplify data interpretation, because weak spectral features can be seen more clearly.

In a practical sense, these two developments mean that Auger analyses can now be performed routinely, easily and rapidly on sample features which are on the order of 1000 Å or less. In terms of spatial resolution, no other existing surface analysis technique can compare to this capability, which is leading the way to numerous new areas of application for Auger analysis.

References

1   Auger, P., Comptes. Rend. 177, 169 (1923)
2   Auger, P., J. Phys. Radium 6, 205 (1925)
3   Harris, L. A., J. Appl. Phys. 39, 1419 (1968)
4   Smith, G. C., "Surface Analysis by Electron Spectroscopy - Measurement and Interpretation," Plenum Press, New York (1994)
5   Linsmeier, C., Vacuum 45, 673-690 (1994)
6   Frank, L. and M. M. El Gomati, Czech. J. Phys. 44, 173-193 (1994)
7   Briggs, D. and M. P. Seah, "Practical Surface Analysis," 2nd ed., Vol. 1 - Auger and X-Ray Photoelectron Spectroscopy, John Wiley & Sons, New York (1990)
8   Ferguson, I. F., "Auger Microprobe Analysis," Adam Hilger (IOP Publishing Ltd.), New York (1989)
9   Joshi, A., "Auger Electron Spectroscopy," in "Metals Handbook," 9th ed., Vol. 10 (Materials Characterization), American Society of Metals, Metals Park, OH (1986)
10 Grant, J. T., Appl. Surf. Sci. 13, 35-62 (1982)
11 Castle, J., Met. Mat. (May), 268-272 (1992)
12 Forsyth, N. M. and S. Bean, Surf. Interface Anal. 22, 338-341 (1994)
13 Narum, D. H., J. Vac. Sci. Technol. B 11, 2487-2492 (1993)

14   PHI 670 Auger Nanoprobe Technical Instruction
     Videotape Guide, Perkin Elmer Corporation,
     Eden Prairie, MN (1992)

15   Rathkey, D., Microscopy Today (June) 93-4, 14
     (1993)

16   Delong, A., Microscopy and Analysis (November),
     USA ed., 17-19 (1993)

17   Goldstein, J. I., D. E. Newbury, P. Echlin,
     D. C. Joy, A. D. Romig, Jr., C. E. Lyman, C. Fiori
     and E. Lifshin (Eds.), "Scanning Electron
     Microscopy and X-Ray Microanalysis," 2nd ed.,
     chapter 2, 21-68, Plenum Press, New York (1992)

18   Lindquist, J., D. Rathkey and P. Fisher, R&D
     Magazine (June), 91-98 (1990)

19   Mizia, E. M., The PHI Interface 13 (1), 1-4, Perkin
     Elmer Corporation, Eden Prairie, MN (1990)

20   Cameca News (May 1992), Cameca Instruments,
     Courbevoie Cedex, France (1992)

21   Stickle, W. F., L. A. LaVanier and D. F. Paul, The
     PHI Interface 13 (3), 1-9, Perkin Elmer
     Corporation, Eden Prairie, MN (1991)

22   Olson, R. R., L. A. LaVanier and D. H. Narum,
     Appl. Surf. Sci. 70/71, 266-272 (1993)

23   Sekine, T., N. Ikea, Y. Nagasawa, M. Kudo,
     M. Kato and S. Hofmann, JEOL News 31E (1),
     41-45, JEOL Ltd., Tokyo, Japan (1994)